CFZ YEARBOOK
2024/5

Copy Edited by Guin Palmer
Typeset by Jonathan Downes,
Cover and Layout by SPiderKaT for CFZ Communications
Using Microsoft Word 2000, Microsoft Publisher 2000, Adobe Photoshop CS.

First published in Great Britain by CFZ Press

CFZ Press
Myrtle Cottage
Woolsery
Bideford
North Devon
EX39 5QR

ISBN: 978-1-909488-70-0

For Maxine

Introduction

Dear Friends,

It is quite a strange feeling to be dictating this introduction to the latest CFZ Yearbook, considering the fact that I have been doing these yearbooks on and off (mostly on) since the dying days of 1995, when my first wife, Alison, and I put together the very first one.

A lot has changed since then, but of course it would have done - it has been 28 years after all. The biggest change happened nearly 20 years ago when the CFZ relocated from Exeter to my old family home in North Devon. At the time of writing (early October), the sad news is that I don't know whether I will be able to stay in this house. Put simply, because of the intransigence of new legislation which has been passed, in the 18 years since I took out this mortgage, I will not be able to get a renewal and will have to move back to Exeter. This is something which I really do not want to do. By the time you read this, I suspect it will all be done and dusted one way or another, and as so often is the case, when I am trying to write deathless prose in advance of the publication date, anything which I am writing today may be completely out of date by the time you read it.

As I expect, many of you will know, I have a parallel career as a music journalist and historian and, as anyone who has ever read my ramblings will also know, I am a particular fan of The Beatles. I have always been interested in their ill-fated record company, Apple, which started out with all sorts of high concepts and altruistic ideas but which ended up with a whole bunch of people ripping them off and bringing them close to bankruptcy. These people, who should have known better, had been considered as trusted friends by the band.

One would have hoped that I should have known better after reading so much on the subject, but like so many people, I did not pay any attention to the lessons of history. Thus, when I inherited a property and a certain amount of money, which would allow me to do the sort of things I wanted to do with the CFZ, I went out and did them. With hindsight, this was both over-trusting and over-generous with some (but certainly not all) of the people to whom I entrusted the CFZ. Over the years, sadly, I have been summarily ripped-off by a whole bunch of these reprobates. Unlike the 'Fab Four', I didn't find that people had nicked the lead off my roof, but I did find that I had lost many thousands of pounds and that various things belonging to both the CFZ and to me personally were stolen. As I am sure you'll agree, this is not something which I am happy about. Sadly, there is nothing I can do. As someone once said, "Trust no one".

Consequently, these unfortunate events, along with the untimely death of my wife a few years ago, not to mention my declining health, have not helped the CFZ to achieve what I have wanted it to. We have still achieved quite a lot nevertheless, and I intend to achieve a lot more still.

The CFZ has now been going for 31 years, and it is my sincere hope that it will continue to do so long after I am quietly laid to rest. Because, what we are doing, I believe, is very important and as the world around us becomes increasingly more insane and makes less and less sense to silly old fogies like me, the importance of our work will be consolidated further I'm certain. Since the rise of the internet and in particular social media, it means that anybody can describe themselves as 'a cryptozoologist'. Although this is precisely what I did at the beginning of my cryptozoological career, and that - indeed - Richard Freeman, Graham Inglis and no doubt several other "heroes of modern Italy" have done so, on the whole we have tried to be sensible, logical and to act within the parameters of the scientific method. But, many of the people who claim to be cryptozoologists in the second decade of the 21st century, are entering into flights of fancy which Bernard Heuvelmans, the 'Father of Cryptozoology' would have had trouble acknowledging. I knew Bernard and he was our Honorary Consulting Editor from 1994 until his death in 2002. Bernard was not hidebound in his thinking but would, I am certain, be absolutely appalled at the current widely accepted paradigm that cryptozoology is a branch of paranormal research rather than anything to do with the natural sciences.

This is manifest nonsense. And whilst, indeed, some of the stuff with which we deal could be categorised well into the areas which are described vulgarly as 'weird shit'. When we do investigate such things we do it as methodically and as sensibly as possible. I personally don't think there is anything unscientific about dealing with a subject like 'The Owlman of Mawnan', if one is not sensible, logical and methodical. If one does not comply with what my old maths teacher at Bideford Grammar School used to admonish me for and 'show one's workings', then one has completely lost the right to be taken seriously.

Something else which I wish to make clear about this yearbook and the CFZ in general, is that there are no CFZ Thought Police. There are no official CFZ paradigms to worship. The Centre for Fortean Zoology is a clearing-house for information and for theories and if one person's theory contradicts the theory of another person, this matters not at all. What does matter is that both sets of theories are well presented, well argued, do not support any single religious or political movement and that the evidence has not been twisted to do so. Apart from that, and a complete ban on anything that promotes animal cruelty there are no rules, and there are people within the CFZ who have a rich and diverse set of beliefs. And as long as they behave in a gentlemanly or ladylike manner, at least in public - and in this sentence the word "public" also includes social media - then they are welcome within the CFZ.

In recent years, I have become interested in the concept of 'Reality Tunnels', as first categorised by the late Robert Anton Wilson and the equally late Timothy Leary. I always use Leary's name with a feeling of trepidation, as he is best known for his experiments with psychedelic drugs and his dictum, "tune in, turn on, drop out". My career as a psychonaut began and ended over 40 years ago, and I have to admit, that I was not particularly impressed by the psychedelic experience. However, Leary was a well known clinical psychologist in the early 1960s and, I for one, find the comfort of 'Reality Tunnels' to be absolutely enthralling.

Basically, everyday our bodies and minds are bombarded with all sorts of diverse stimuli, far more than our pitiful central nervous systems can effectively manage. And so, as individuals, the nervous system of each human being has to "decide" which of these different stimuli they are going to recognise and consequently react to.

It is a mildly annoying phrase in the English language that human beings have "five senses". Actually, without going into the areas of pure "esoterica", we have dozens. For example; as well as sight, sound, touch, taste and smell, we have the ability to feel heat, cold, pain, atmospheric pressure, gravity fear, the effect of endorphins and a host of other things. I once tried to make a list of them all, one night, not counting anything that is usually lumped into the "paranormal" and I gave up just before I reached 100. I can seriously recommend this as a useful exercise to help you get to sleep.

And so, if each of us experiences a unique mixture of stimuli, then this would explain why, in so many cases, people who experience something out of the ordinary may have a noticeably different experience, to somebody who is standing next to them. To give an example, in January 2003, I was at Bolam Lake in Northumberland, together with several CFZ people and a contingent from a local paranormal research group. Over the previous couple of months, this lake had been the location where a whole string of reports of a "bigfoot' type creature had been made. I have told the story of what happened that evening on many occasions. However, it is interesting to note that when three or four other witnesses and I saw "the creature", several of us had different experiences to others. Especially if my theories about magnetic anomalies in the area are true (magnetism being yet another of those geophysical stimuli which we can experience), then the reality of Reality Tunnels make far more sense, than any of the bits of paranormal hocus pocus which, sadly, so many people seem to believe.

Taking a brief aside. I dislike the use of the words 'paranormal' and 'supernatural'. Because I believe they are perfectly normal and perfectly natural.. It is simply that they are defined by the laws of physics that we do not yet understand. And as somebody who failed his physics 'O' level in 1976, my knowledge of such things has not improved……so I shall not attempt to explain them here.

The natural world is still largely misunderstood, and there are fortean zoological investigations which need to be undertaken all over the world. There are even investigations that younger people can do from home. The CFZ exists because we believe that these anomalies of the natural world are worth studying and that they may tell us all sorts of things which could well affect our own future on this planet.

So, I am inviting you all, (those of you who are not doing so already) to join the hunt and become more closely involved with the Centre for Fortean Zoology in our quest to make more sense of our universe.

Yours, as ever,

Jon Downes
(Director, CFZ)

Faculty of the Centre for Fortean Zoology

Contents

WOODPECKER.

The strange "horn" / Photo: Hans-Jörg Vogel,

Findings, Analyses, and Opinions on a Strange "Horn" from Cambodia

Hans-Jörg Vogel / Berlin

A few years ago, our association in Germany, then Interessengemeinschaft Kryptozoologische Forschungen / IGKF), was sent an unusual horn for assessment. To this day, the exact nature of this object has remained uncertain. However, this article makes a considerable contribution to its clarification.

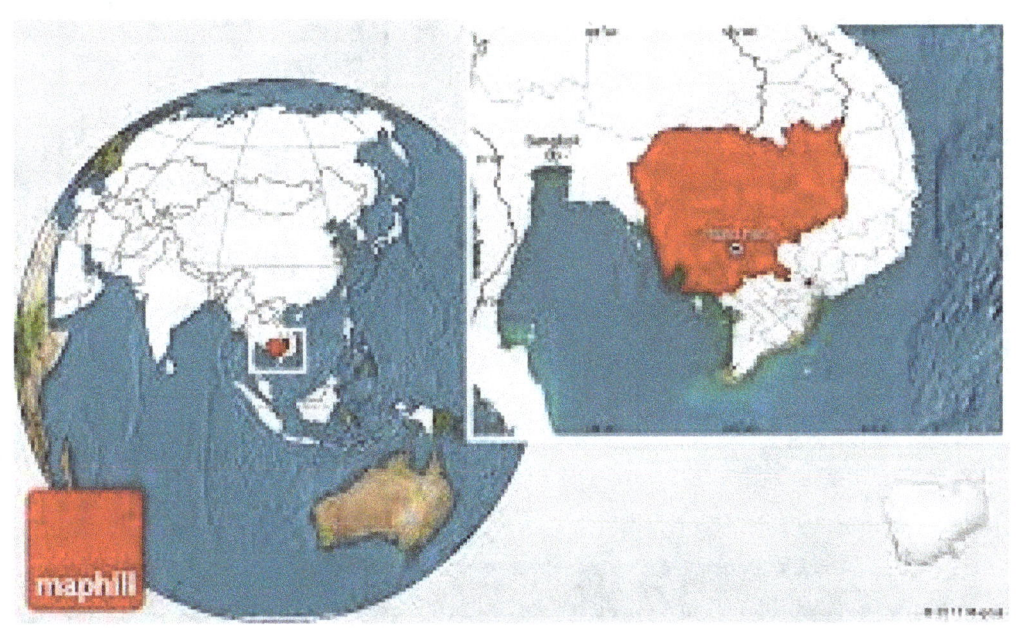

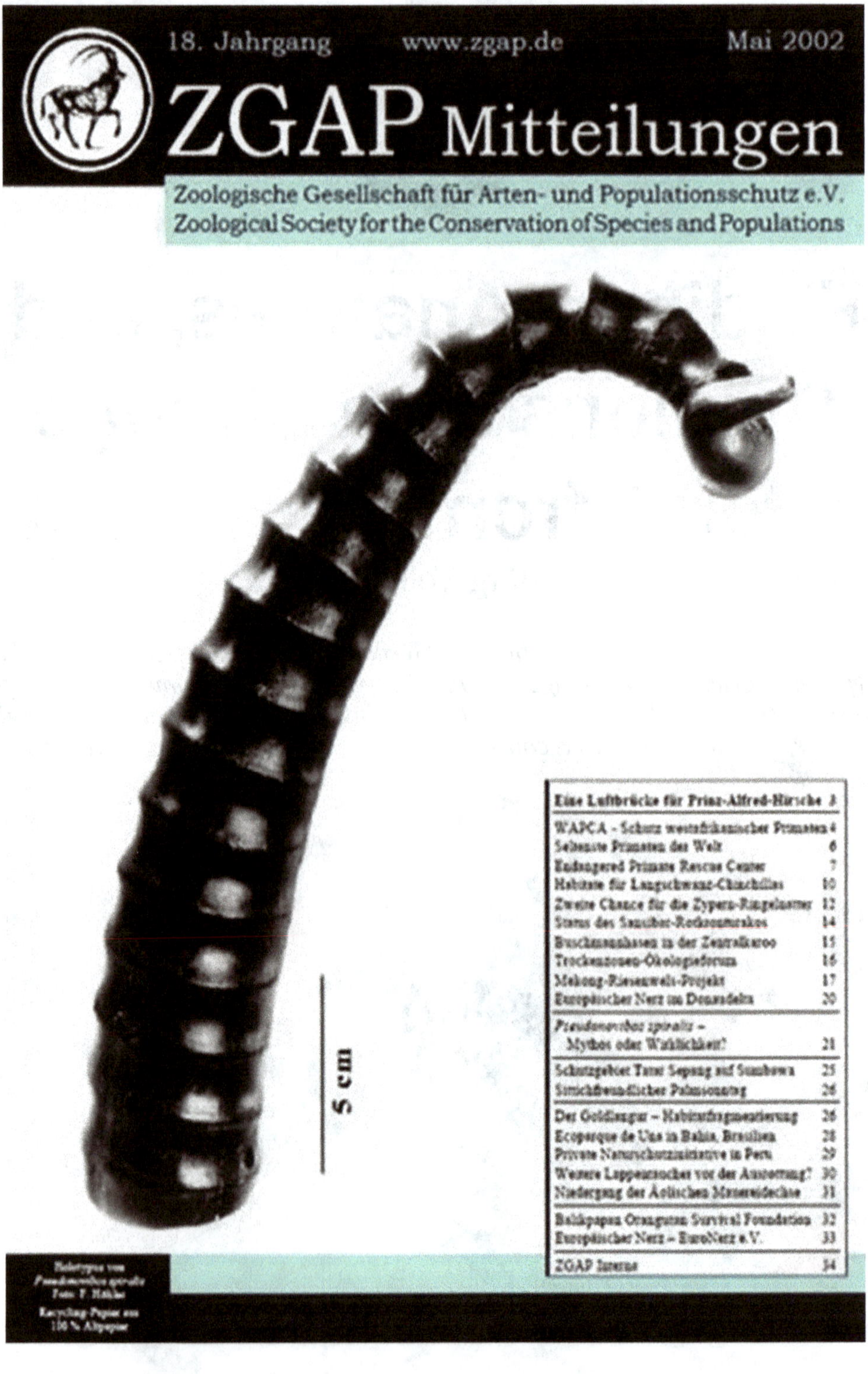

18. Jahrgang   www.zgap.de   Mai 2002

ZGAP Mitteilungen

Zoologische Gesellschaft für Arten- und Populationsschutz e.V.
Zoological Society for the Conservation of Species and Populations

5 cm

Holotypus von
Pseudonovibos spiralis
Foto: F. Höhler

Recycling-Papier aus
100 % Altpapier

The horn of a *pseudonovibos spiralis*?, holotype, photo: F. Höhler (2) Quelle:
https://www.zgap.de/

After an event held by our organisation in September 2003, in Berlin, which was covered in the German media, we were sent a small package containing an unusual object. Its sender (Mr. S.) described it as a possible horn of what appeared to be a yet unknown animal species. Mr. S., the owner of an antiquities shop, had brought the item home from one of his many travels abroad. An article about our event prompted him to contact us and to request our assistance with its identification. Eager to help, and equally curious, we launched an immediate investigation into the "horn".

Mr. S. indicated the country of origin of this horn was Cambodia. Initial research led us to look into the mythical *kting voar (pseudonovibos spiralis)*, or "snake-eating cow," a possibility we discarded quickly given shape and size of the object. [1]

The horn of a *pseudonovibos spiralis*?, holotype, photo: F. Höhler (2) Quelle: https://www.zgap.de/

The next phase of our research entailed contacting zoos, wildlife parks, museums, and other institutions, to obtain expert opinions on a photo of the horn. Unfortunately, we received few responses, among them, however, from a number of renowned institutions, namely the State Museum of Natural History in Görlitz, the State Museum of Zoology in Dresden, the Museum of Asian Art / Berlin State Museum, as well as the Museum of Natural Sciences in Berlin. They all agreed that this "horn" cannot actually be attributed to any currently known species of animal.

Subsequently, we were successful in arranging for an appointment with a preparator at the Berlin Museum of Natural Sciences, to conduct an examination of the "horn". The preparator was able to confirm that the object was indeed made of the same substance as an animal horn, but could not attribute it to any known animal species.

Upon closer examination, he noticed several areas on the surface of the "horn" that indicated signs of manipulation. These particular areas had already caught my attention as well. Could it be an artistically crafted object after all? If so, the question arose as to which known species of animal it could possibly originate from.

Below detailed photos and measurements:

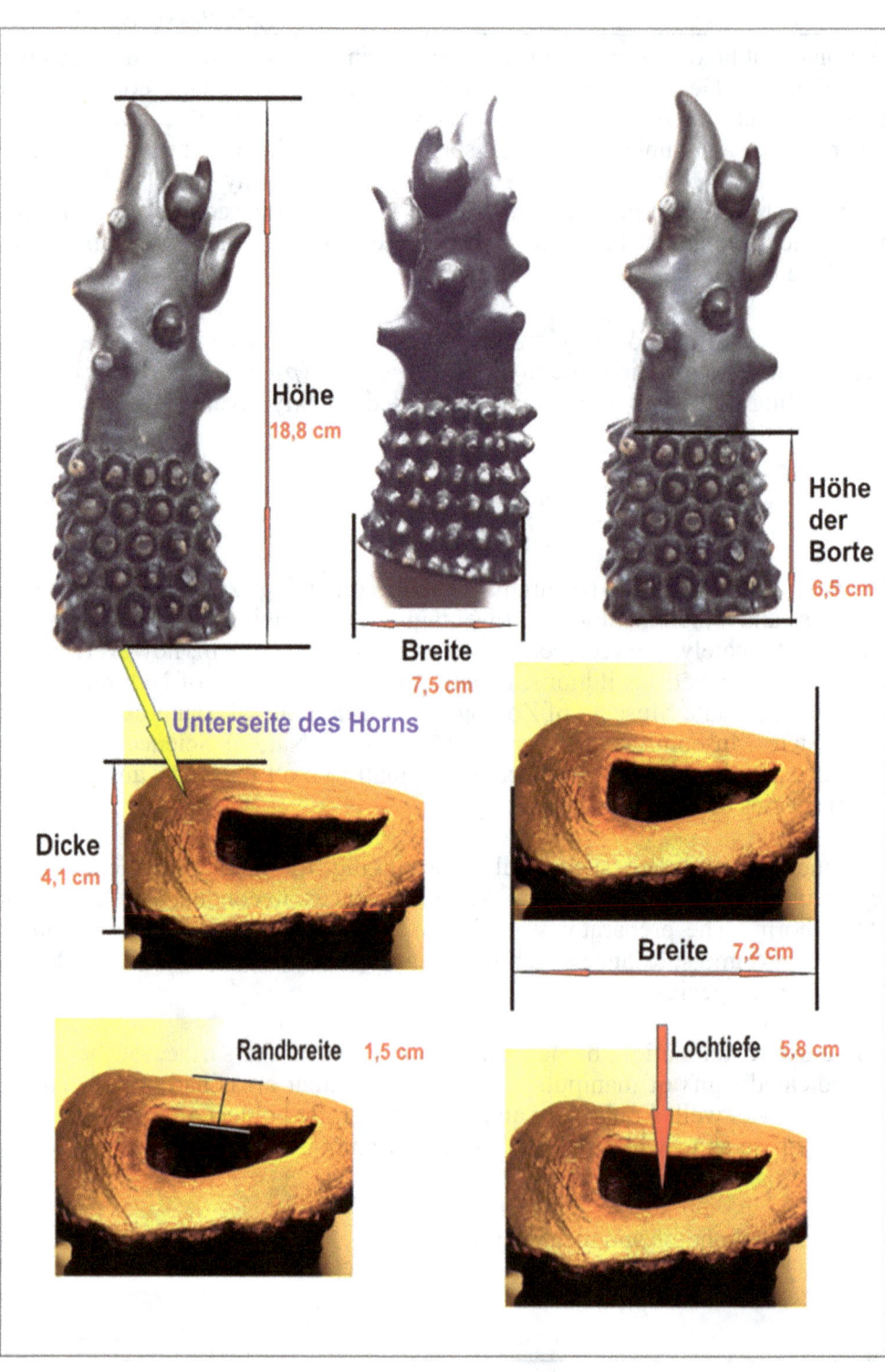

Höhe
18,8 cm

Höhe der Borte
6,5 cm

Breite
7,5 cm

Unterseite des Horns

Dicke
4,1 cm

Breite 7,2 cm

Randbreite 1,5 cm

Lochtiefe 5,8 cm

In the meantime, Mr. T. N., an expert from Vietnam, responded to my inquiry via email. Mr. T.N. informed me that such horns or similar ones are commonly found in souvenir shops, occasionally in markets or in the pet trade in Vietnam and Cambodia. These horns are derived from water buffaloes and are embellished through carving, to create something unique. The situation is comparable to that of *Pseudonovibos spiralis*, of which only artificially deformed horns have been reported thus far. As per Mr. T.N.'s statement, it remains entirely unclear whether an animal with such horns ever existed. It was however evident in the case of our horn, he continued, that there was no such animal with these horns, but rather artists who had found a source of income through the production of such objects.

As the next step, I requested the opinion of Mr. M.S . He kindly offered to conduct an examination, and we gladly accepted his offer. After some time, Mr. M.S. shared the findings of his investigation with us. What follows is a brief excerpt from the results of Mr. M.S.'s examination:

- Hypothesis: cultural forgery, created skilfully in large quantities, serving both as an object of reverence and as a commodity for tourist trade.
- The dark coloration of the object: This typically indicates artificial manipulation by human hands and is commonly employed in cultural art pieces, whereas natural horns or antlers tend to exhibit lighter shades. However, this coloration could also be attributed to other factors.
- A preliminary assessment based on the material composition was challenging due to the obvious application of a coloured coating, rendering the object relatively smooth. However, no significant signs of rough manipulation could be observed.
- The upper narrow section of the object is densely filled with a material that is not immediately identifiable upon initial inspection. Similar to the exterior, the interior is also coated with a black film, suggesting that the object was immersed in a liquid substance.
- Mr. M.S. had extracted three small samples from the horn for analysis, taken from inconspicuous areas. First, a scraped sample of the black coating was obtained to determine its origin. Second, a piece of material was scraped off the lower edge. The third sample was taken from the solid filling material of the object through the lower opening.
- Upon examination of these samples under ultraviolet (UV) light, tiny grooves became visible, suggesting scrape marks. It is possible that these were caused by fine sandpaper or fine steel wool. However, these grooves could have also formed during transportation, when the object may have been carelessly pushed across a rough surface.

- The traces on the round protuberances were even more distinct. The do not appear to be naturally grown, as there is no truly smooth edge present. Some of the protuberances are very sharp at the points where they emerge from the surface. When viewed under UV light, they seem to be intricately carved into the material.
- An examination of the three extracted samples under a microscope reveals that all three are made of fibrous material, while the sample scraped off the upper surface was adhered by a particularly thick layer of the black coating.
- Evaluation of the chemical fingerprint under the IR spectrometer: The coating material is wax with a probability of 97.21 %, predominantly derived from shoe cream or shoe polish. The main components consist of over 60 % lanolin and carnauba wax. Additional components include a wax emulsifier (2 %) used for solution binding, and various oil-based solvents, primarily residues of petroleum.
- The sample from the lower edge showed a high carbon content. The database provided an 84 % match to a wood product and a notable 73 % match to a collagen factor. Collagens are closely related to keratins, which constitute the main component of most animal horns. Therefore, it is clear that this object is either made of wood or remnants of a former animal antler. In any case, it was previously composed of living matter. In the

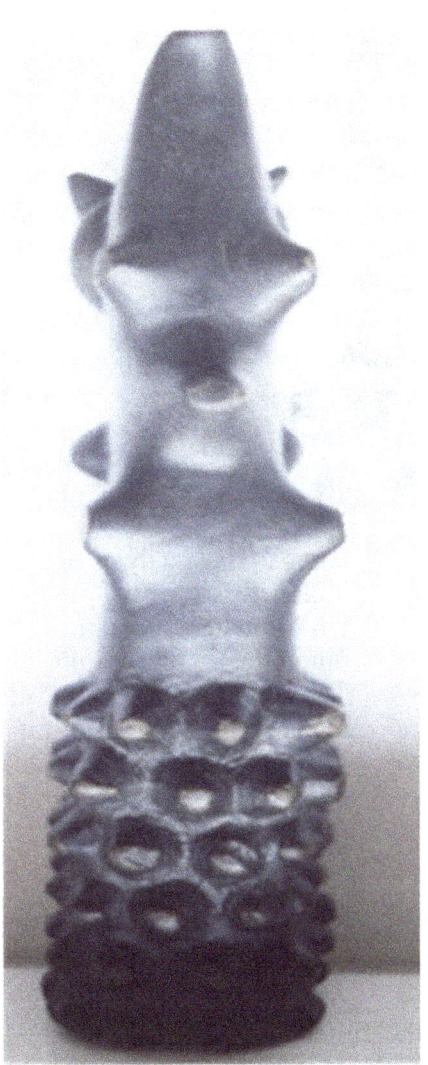

Seltsames Horn aus Kambodscha

Artikel zu den bisherigen Ergebnissen im PDT 19 (2004) veröffentlicht!

comparative analysis against my four selected reference samples, the highest similarity of 56.45 % was found with the deer antler. Unfortunately, contamination from the coating material could not be ruled out in this small scraping sample.

- The composition of the filling material in the upper part of the horn matched the deer antler reference sample to 89.62 %. This provides evidence that, at least in its original form, the horn must have been an antler from a large mammal.

In conclusion, as Mr. M. S. states, it can be established that this object is a beautifully carved object made from a piece of deer antler. However, the precise reason for this particular shape, and whether it serves as a tourist souvenir remain elusive. Presumably, it could be a fetish object for various ritual occasions, most likely a stylized phallus used in fertility rituals. You can read the full analysis of Mr. M. S. in the magazine PTERODACTYLUS 19 (II-2004), no. 66–9 (German).

Another expert, Mr. D. D., who unfortunately passed away, also examined this "horn" several years ago and came to the conclusion that it was probably a cult object. He drew comparisons to similar items from Africa. (4) His full assessment and interpretation can also be found in the magazine PTERODACTYLUS 19 (II-2004), no. 6, 10–13.

In summary, it appears that this "horn" from the Asian region is most likely an intricately carved work of art, tourist souvenir, or cult object. The question that remains unanswered is the specific animal from which it originated before undergoing this artistic transformation.

References and Further Reading:

1. <https://www.spektrum.de/lexikon/biologie/pseudonovibos-spiralis/54572> (27 May 2023).
2. ZGAP Mitteilungen (Mai 2002), no. 18 (www.zagp.de).
3. Cryptozoological Magazine PTERODACTYLUS 19 (II-2004).
4. Heike Owusu, Symbole Afrikas (Schirner Verlag: Darmstadt 1998), 17.

All individuals mentioned in this article by initials may be contacted upon request. Please send any questions to <archiv-afrikanische-kryptide@gmx.de> and we will forward them. Moreover, interested readers may also request the relevant article from the magazine PTERODACTYLUS 19 (II-2004), in PDF format (German only) using the same email address.

Jana Holland translated the article from German into English.

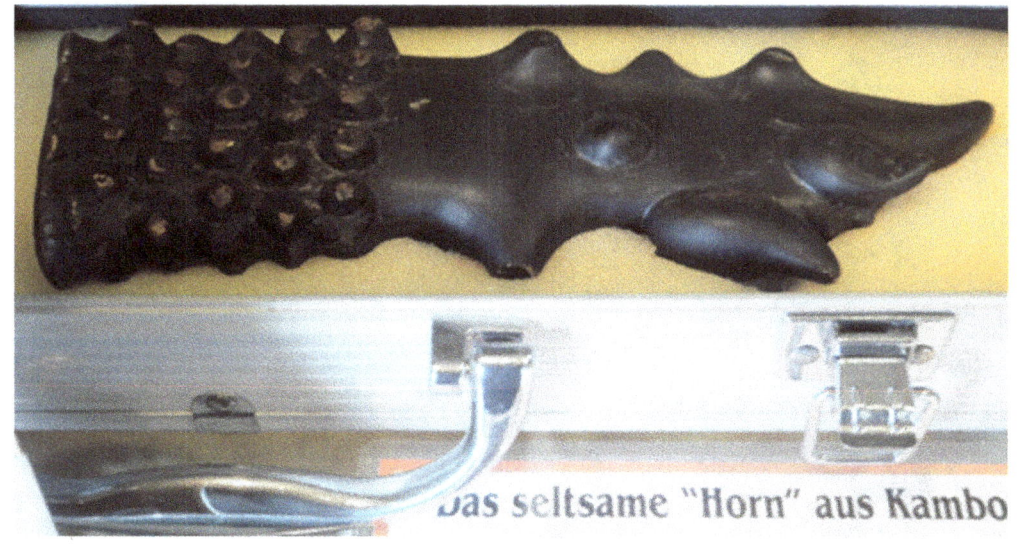

Das seltsame "Horn" aus Kambo

The "horn" is part of the exhibition at the Museum Tor zur Urzeit e. V. in Brügge (Schleswig-Holstein).

www.torzururzeit.de

True believers see more patterns

Michela de Mattei

A full account of the 2015 Tasmanian tiger sighting in southern Tasmania can be read below. This information was taken from personal notes on the thylacine, provided directly by witness Greg Booth in 2016, to Adrian Richardson. This sighting has long been told only in words, it travelled orally from one person to another, then in this empathetic channel has finally found its written publication.

I believe it should be passed on to the cryptozoological community exactly as originally chronicled by Adrian Richardson. Greg Booth, the man who had the fortunate encounter with the thylacine, has sadly disappeared, no one can track him down and I like to imagine that he is hiding in the wild with our elusive creature: the thylacine.

My contribution will consist of a personal digression, that will follow the report also illustrated by some pictures - to be read as an additional track to trace the animal. I believe that digression, by its very form, can help the imagination to operate, expanding the narrative and forcing us to deviate and reinterpret the signs. Certainly everything would be solved if we had a sense of smell.

The Greg Booth Tasmanian Tiger Sighting
3 April 2015

Date Friday 3rd April 2015 **Time** 1:20 pm
Location Forestry track SW Tasmania **GR** 6 - - 9 - -
Weather condition Mild & 17 degrees C
Persons present Greg Booth   George (Joe) Booth
               (Woodcutter)   (Retired Forester)

The Reason of being in the location

Greg and Joe were having some valued father & son time together exploring a Tasmanian State forest. Joe was showing his son areas of bushland where he worked as a forester many years prior.

Joe had remembered an overgrown track in the area that had been constructed during the Tasmanian early settlement days. The Dawson Track had been cut in and worked upon by convict labour that had to endure the tough terrain including the harsh penal conditions from many years ago. It was to be the last stop in the area before returning back to Joe's home town of Ellendale.

Greg was driving for his dad in his father's Toyota single cab 4WD ute and was instructed to drive some 400 metres up a secondary forestry track & park the vehicle at a Y junction.

Background

In early April of 2015 I was notified by a friend from Wayatinah, a small town in the Tasmanian Central Highlands that an acquaintance of his had a Tasmanian tiger sighting.

In the same conversation my friend informed me that the family involved were no longer talking to anyone about the sighting. Even though I was given a surname to follow up, I took my friends advice and decided to leave the family alone.

The following month in May, Joe Booth an elderly retired forester who was present at the sighting, had written his record of events in (his own words) into The Highland Digest (Issue No 182 of May 2015, page 9).

A footnote to Joe's report indicated that another Tasmanian tiger sighting had taken place only 3 kilometres away in the same area by a Log truck driver. As it turns out this separate incident is an interesting story from 2012 and that the Log truck driver (name withheld) had to stop his truck in fear of running over the animal.

Upon reading the Digest article written by Joe, I instantly felt from my previous thylacine research experience that this was a Tasmanian tiger sighting like no other & that there was much more to this story than what was written into the monthly local community newsletter. (Joe was 79 years old at the time of the sighting & of writing the article)

Indeed there was more to the story, and at the time I did write into my thylacine journal notes that this encounter would awaken the scientific world. However this never happened as this sighting was not reported to the two major local newspapers or media outlets.

The Tasmanian tiger sighting was only spread by word of mouth throughout the local rural community. By the 3rd of June 2015 (two months to the day after the initial sighting) I had completed 6 reconnaissance trips into the area trying to locate what I thought could have been the likely location of the sighting. The only bearings I had was the Dawson Road and this involved me having to drive some 1,400 kilometres during my research. It was my 6th and final trip (and taking photo's) that I ended up only 50 metres shy of the actual sighting position which was indicated later when

Greg & I were on our research trips in the area together. Greg was amazed by looking at the photographs that I had taken from my previous trips that I had stumbled into the actual position without a guide & only the use of 1:50,000 conventional maps.

It wasn't until the following year in 2016 that I received a surprise phone call that came out of the blue from Greg Booth, asking if I would assist him with his own private search.

When Greg made this initial phone call little did we know that we had met some 18 years prior as he was a partner to my wife's niece.

In hindsight really I should have made contact with Joe & Greg in April 2015 however at that time I did not relate that Joe was Greg's father.

Greg needed to be heard reference his sighting. He expressed in his (own words) the sighting & of the countless times that he had reflections & images mirrored back to him from his flashbacks.

Greg was very sincere and emotional when describing the animal he saw.

Unfortunately Greg found some people within his rural local community had doubts that he had seen a Tasmanian tiger. (Like most of the general population, due to the fact that we have all been educated to believe that the Tasmanian tiger is extinct)

Once again ridicule raised its ugly head just like it does with majority of Tasmanian tiger sightings here in Tasmania. The common phrase I hear is "don't you know that the tiger is extinct".

Ridicule is why many sightings are not reported to Tasmanian authorities.

Greg to his credit did not in any way waiver from his original statement, including when he actually did a walk through on location of the tiger sighting with leading Tasmanian biologist Nick Mooney.

Greg stated in 2016 that he thinks of that tiger he saw at every waking moment and that he still gets flashbacks of the sighting.

Two years on from his sighting (in 2017) Greg still couldn't get his unprescendented tiger sighting out of his mind.

To give reflection to what Greg saw I believe at this point would be an ideal time to mention a quote by Brene Brown, an American professor & author.

"In order to empathize with someone's experience you must be willing to believe them as they see it, and not how you imagine their experience to be".

The Tasmanian Tiger Appeared from No Where

As previously explained Joe's vehicle was parked at the Y junction.

The vegetation covering the area was rather thick with Tea tree, young & old eucalypt trees plus other species of trees including Myrtle & Celery Top Pine.

This area had been logged several times before over many years by Forestry Tasmania (now named Sustainable Timber).

Greg was instructed by his father to walk & follow the forestry track back to where they had just driven while Joe would venture some 20 metres into the undergrowth to try and follow the old convict track.

Impressively for a man of Joe's age he is very active & could skilfully jump over logs & walk the thick undergrowth with ease.

Greg was now moving down the forestry track and keeping in line with his father who was mostly unseen but could be heard from within the undergrowth following the old abandoned convict labour built track.

Having walked some 400 metres down the track and around a bend Greg decided to stop & wait on the edge of the gravel & wait for his father to catch up with him. At this point of time Greg was thinking of rolling a cigarette & he was just staring & looking into the undergrowth. Directly in front of him the bush was rather thick & there was two large pushed or fallen eucalypt trees lying across the hidden convict track.

Greg was just staring into the undergrowth when his world of normality changed.

As Greg said, it was then that everything suddenly happened.

Out of the thick bush from under the closest fallen eucalypt tree came an animal and it walked directly towards him.

The animal did not notice Greg until it got within approximately 3 metres or 10 feet from him.

Both Greg and the animal were in shock & disbelief in meeting each other. (Greg told me that he was totally gobsmacked with this initial encounter & it was an animal that he had not seen before ever, in his time in the bush)

It was at this point Greg stated that he realised he was looking at a Tasmanian tiger.

On seeing Greg the Tasmanian tiger went into the seated position.

The seated tiger & Greg just stared at each other. (One could not imagine what was running through Greg's mind at this stage)

Greg estimated that the time spent staring at each other would have been 5-6 seconds in duration. This is the reason how Greg managed to highlight the facial features of the animal in fine detail, including a scar above its right eye.

The description of the thylacine given by Greg was rather extensive which included its dark & deep set eyes, with white patches below each eye with a very long snout, dark nose with a strong jaw line, short ears with a tough thick neck.

In his own words as he was describing the thylacine Greg said that the tiger was not a pretty animal to look at and it wasn't in as good condition as his own domestic dog. (Greg's dog being a solid well fed tan coloured French Boxer)

The colour of its fur was short and not thick being no longer than 1cm or ½ inch in length and it was light tan.

Once the initial shock of the meeting ended the seated tiger used its tail to assist itself to all four feet & pushed itself upwards some 70-75 cm prior to landing onto its front feet, just like a kangaroo would when it uses its tail to move.

Greg noted that the front legs looked shorter than the rear legs.

Once the Tasmanian tiger was standing on all four feet it turned to its right & started to walk onto the very edge of the gravel & move back up the forestry track heading towards the parked vehicle. Greg stated that the Tasmanian tiger stood at a man's knee height of approximately 50-60 cm (2 feet).

It was at this point Greg saw the rear opening pouch and the enlarged bottom area including the exposed anus region of the animal.

Greg explained that the rear opening pouch was approximately 18 cm or 6 inches long, or about the length and size of a man's open hand. It reminded him of seeing a Tasmanian Devils rear opening pouch.

The tiger had approximately 15 dark stripes running along its back and very little hair on its tail but there was a tuft of fur at the very end, and that the tiger's tail was stiff & straight with a light kick up at the very end. In fact Greg stated that the tail was so straight you could put a builder's spirit level on it and the vertebrae could be seen along the length of the tail.

At some point Joe called out to Greg but he couldn't respond properly as he was trying to keep up with the now trotting Tasmanian tiger that was moving back up the track & heading towards the parked vehicle.

Greg couldn't see the legs of the tiger at times due to the bush however he could see the top of the tiger as it moved in & out of the bush and onto the edge of the gravel track.

Greg believes at this point while following the tiger he called for his father to get onto the track and follow him.

The two distinctive points that Greg raised was when the tiger was trotting & moving up the track was how the dark stripes disappeared & blended into her light tan fur, and also how the animal maintained its straight & stiff tail when moving. (Further discussions regarding the animal's movements with Greg and he felt that the tiger was wanted to cross the track however refusing to cross due to Greg being in close proximity)

Greg then watched the tiger go around the parked vehicle at the Y junction & then disappear into the Tea tree bushes heading into a westerly direction.

Just as the tiger disappeared into the Tea tree Joe finally caught up and reached his son at the vehicle. Greg then explained what he had just seen to his father.

Unfortunately Joe had not seen the Tasmanian tiger that he had disturbed & flushed out from the undergrowth on the old convict track, however he instinctively knew that it was a tiger that his son had seen & described to him. Joe then asked Greg was it a Tasmanian tiger that you just saw? Greg replied immediately with a definite yes!

Greg said that prior to that day he had never really given any thought to the Tasmanian tiger & that he had believed the animal was extinct. (For someone that thought that the thylacine was extinct then Greg's description & fine details were astonishing)

Greg's ability to observe & then describe (in his own words) were amazingly skilful. I had witnessed his observation skills on many occasions during our bush trips together when placing out game trail cameras.

Conclusion

One of the biggest problems Greg had following the sighting was trying to convince the nonbelievers that he actually stood only 3 metres or (10 feet) from the animal.

Greg only wanted others to believe in what he had witnessed on that day. Greg had nothing to gain but plenty to lose as his personal credibility was at stake. On occasions Greg would say why me? And how come it happened to me?

My reply to Greg was you were fortunate to have that close encounter and that statistically it is about the same as winning Lotto, as that is about 1 in 4 million chance of winning. You actually won Lotto except you didn't win the money.

People have asked Greg as to why he didn't get a photograph of the tiger due to the length of his sighting? Greg was not carrying his mobile phone with him as it was on the seat in the vehicle parked at the Y junction. There is no mobile phone reception in the area so it was left in the vehicle. Greg also stated that he wouldn't have even thought of taking photographs with his mobile phone because the event was just so surprising and very quick. Greg also rightly stated, who walks around out in the Tasmanian bush with a mobile phone in hand?

Greg stated that he saw the Tasmanian tiger a total of 3 times and each sighting being 5-6 seconds duration.

Having walked this 400 metre track many times I believe that Greg may have under estimated his time when watching & following the thylacine moving in & out of the Tea tree scrub alongside the forestry track.

Nick Mooney a leading Tasmanian biologist on the thylacine who went to the site mentioned to Greg and I that he too believed that the complete sighting time combined while watching the moving tiger would have been longer than the 6 seconds that was estimated by Greg.

It is to be noted that Greg & Joe did not notify or report the tiger sighting to any local Tasmanian newspapers, or media outlets, or the Tasmanian Government Department of Primary Industries, Parks, Water & Environment (DPIPWE).

The Greg Booth Tasmanian tiger sighting was spread only by word of mouth and the written article by Joe in May 2015 issue of The Highland Digest.

The reason that I have documented the Greg Booth Tasmanian tiger sighting from 2015 is because if it wasn't recorded directly from my witness notes that were taken in 2016 then this Tasmanian tiger sighting could possibly be lost forever.

Adrian Richardson
(Thylacine Research)
22 July 2020

EXIT

We think too much in terms of history, whether personal or universal. Beginnings belong to geography, they are orientations, directions, entries and exits[1]. The Thylacine, a.k.a. the Tasmanian tiger, apparently exited history in 1936 only to re-enter it not long after. Since 7 September 1936, it has actually entered and exited many times:

IN whenever someone claimed to have spotted it in the wild

OUT whenever someone else stated that it was a hoax

Exit is a technology-related concept, highly popular these days since S.Balaji wrote *The Network State*, a controversial book in which he envisions a society that can reduce the influence of bad politics on people's lives by building opt-in societies, places where individuals are free to enter and exit peacefully, run by technology.

The fundamental idea behind the Network State, is to assemble a digital community and organise it to crowdfund physical territory. 'The main problem is that everything, including land, is turned over to techno-capital. A network state in practice can only be neo-colonialist, as significant amounts of capital would need to be raised to purchase enough land to give investors the confidence to pressure existing governments to concede to their desires. In all likelihood, land would be purchased in poor, previously colonized countries that can be more easily swayed by foreign capital. [2]

Putting aside these dangerous techno-capitalist theories, it seemed tempting and fertile for me to consider digital space as a parallel land where Tasmanian tigers still dwell. My search for the Thylacine thus began and developed in this de-materialised and highly populated land: the web. The question of its actual survival is very much debated within online communities, but for me it was no longer a question of whether the animal exists or not. As far as I am concerned, the acts and effects that the Thylacine generates testify to its existence. My conviction since the start was 'never question its existence, but instead understand how it exists'. Rather, follow the path of the animal in

the lives of the beings it connects. Tracking the living beings and the Thylacine in what holds them together. I therefore began to create my own cartography of these relationships. It was necessary to make room for this animal that we had already once forgotten in the cold. The thylacine allowed me to enter its unique and geographically decentralised territory, showing me parts of Tasmania, virtually annexed to some hidden corners of Australia or Papua Guinea. It guided me through the techno-diaspora of the many users participating in the Quest, through Discord servers, Facebook groups, IG pages, database lists, UK amateurs or passionate Sicilian students. The animal seems to peregrinate in many people's lives. In particular, I have been interested to hear the accounts of those who have been lucky enough to encounter it in person. The abundance of sightings certainly proves that we cannot cope with the Tasmanian tiger's demise, but it also made me realise that if the animal haunts us, it is vital to collect these visions and listen to what the animal has to say.

TO SLEEP FACE UP

Juanicu warned me, "Sleep face up! If a jaguar comes he'll see you can look back at him and he won't bother you. If you sleep face down he'll think you're aicha (prey; lit., "meat" in Quichua] and he'll attack." If, Juanicu was saying, a jaguar sees you as a being capable of looking back - a self like himself, a you, he'll leave you alone. But if he should come to see you as prey - an it - you may well become dead meat.[3]

We must be ready to look back, because the moment may come at any moment when an animal appears and stares at us. But what if this animal is an alien animal, or a creature that supposedly no longer exists? What kind of look will it give us and how will we see it? When I started collecting sightings of thylacines, I realised that they all had a similar tone, the accurate account mixed with dreamlike details. Probably, since the thylacine is not a creature whose existence is recognised, the event of a sighting falls into the unofficial domain, making the character of all reports fall into an intimate and almost secret sphere. Witnesses, when given the opportunity to tell their story, add their innermost feelings and impressions. The animal in the course of their stories acquires a mysterious aura, and often takes on an increasingly hybrid form being compared to many other animals: it has a tail similar to that of a kangaroo, a dog-like stature but it is slightly more similar to a fox, and it also has the gait of a horse. The descriptions I received in my inbox shaped in my head a multifaceted creature, made up of different individuals and altogether quite graceful. Certainly of the order of things I have never seen and would love to see. But I think what Juanicu reminds us is that the way animals see us matters. Other kinds of beings - when standing in front of us - they get an idea of us, their own representation of who we are. And what might a Tasmanian tiger - declared extinct - think if it finds us face down?

There is a kind of 'pride' in sleeping looking up, as if we want to show our unconscious side and our vulnerability. Without the fear of having our eyes closed and not being in control. Being ready with our eyes closed. As if at any moment another

being could arrive to share our most submerged part: the nocturnal flow of dreams.

In many cultures, sleep is not the solitary, consolidated eight-hour sleep that Western societies are used to. In some cultures, such as the Runa of Ecuador, sleep is a collective experience, largely exposed to the outside world and shared with others, humans, animals and plants. The night is fragmented, one wakes up in the middle of it to sit by the fire listening to the distant sounds of the forest or the strange dream of someone sitting across the fire. Sleep is segmented so that the dream world can spill over into the waking world. Practising and cultivating this nocturnal state allows one to experience what is usually 'impossible' and to tune in to a world beyond the human. It is true that in a world that leaves little room for shadows, certain things can no longer be perceived, discovered or participated in. When the English colonisers arrived in Tasmania they were hit by the paradisiacal vision of this untouched natural landscape. A novel by M.Kneale tells the story of 20 English explorers who went to Tasmania on an expedition, firmly convinced that the true Biblical Garden of Eden was located there.

This fantasy doesn't really match up with the reverse view, that of the Tasmanians. 'There is some evidence that Aborigines believed the white men were returned ancestor spirits. John West asserted that when Tasmanians died, they expected to re-appear as white men on an island in the Straits'. Robinson's observations also support this claim, he recorded on several occasions that Tasmanians went to England when they died. "An Aborigine born around 1820, wrote personally of this belief in later life, claiming: 'black people

Cradle Mt. 5,068.'

died then arose from the dead [and] became white men'."[4]

How others see us matters. One cannot imagine what went through the mind of Greg Booth - the witness of the report transcribed here - in those five to six seconds when he and the sitting tiger stared at each other. However, I would be far more interested to know what was going through the head of the Tasmanian tiger.

NO ONE SHOULD EVER BE LEFT ALONE WITH A VISION

The fact that the thylacine returns and reveals itself to hundreds of people is a reality, then we can establish that there are many degrees of reality and investigate to which degree each sighting belongs. Since 1936, there have been many reports of this animal resurfacing as if it wanted to reclaim its habitat and re-include us in it. Because if Tasmanian tigers still survive somewhere in the wild, we humans, have in any case currently been cancelled.

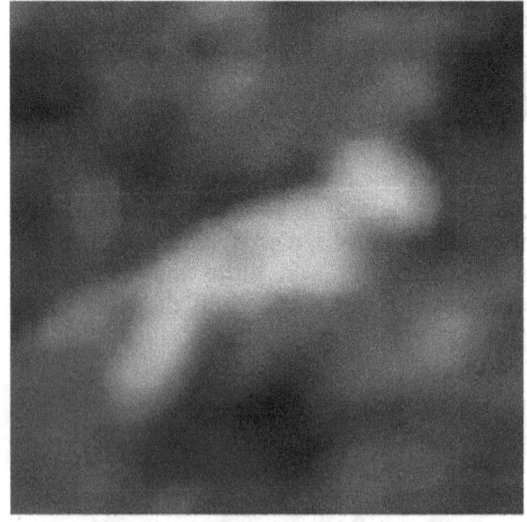

Perhaps a few trees, many leeches, a few Tasmanian devils and a number of bacteria often encounter thylacines and continue to co-evolve with them. We co-evolve with its extinction, with its great lack. There is a tendency to mystify extinction, paraphrasing Van Dooren's words, by making it coincide with the last individual of a species: as in the case of Benjamin, the last Tasmanian tiger. As if, we witness the extinction event at the end, when the last individual dies: Benjamin's demise would be equivalent to the death of a species. But long before that last death, all the relationships in which that animal was involved, both ecological and human, with the communities for which it provided meaning, broke down. This is the dullness of extinction: it is a long, slow decay of ways of life that occurs long before the death of the last individual. In the case of TT, extinction began when man decided to decimate the entire tiger population, first ruining their reputation and then hunting them to the end. The variety of feelings, stories and cultural meanings that bound humans to thylacines gradually faded away.

This makes me think how valuable the hundreds of stories of sightings, evidence, descriptions and debates are. Every observation, every glimpse should be shared collectively to re-establish a relationship with the animal on all levels and recreate its place in our lives. Tasmanian tiger's affairs must be experienced, opposed, measured, enunciated, performed and narrated in various ways, to restore its rich

ecological and cultural role. Is this perhaps the only way de-extinction can possibly occur?

Perhaps only then will the Tasmanian tiger decide it has everything it needs to 'opt-in' again.

REFERENCES
1. G. Deleuze and C. Parnet, Dialogues. Columbia University Press, (1987).
2. Joshua Dávila, "Fork Your Society, I want Out" - Outland Art. [Accessed 16 march 2023.]https://outland.art/network-state-review-balaji-srinivasan.
3. E. Kohn, How Forests Think: Toward an Anthropology Beyond the Human. University of California Press, (2013).
4. N. Clements, The Black War: Fear, Sex and Resistance in Tasmania, University of Queensland Press, (2014).
5. B. Wray, The Rise of the Necro fauna, The Science, Ethics, and Risks of De-Extinction, p.239,Greystone Books, (2017).

1. The bend in the section of the forestry track near the Y junction where the Tasmanian tiger disappeared into the bush in 2015.

2. Greg Booth standing at the position of where the Tasmanian tiger appeared from nowhere directly out from the undergrowth. The tiger then proceeded to walk towards Greg only stopping 3 metres (10feet) from him in 2015.

3. Joe and Greg Booth on the convict track and at the fallen eucalypt tree where the Tasmanian tiger first appeared in 2015.

4. Greg Booth walking on the forestry track where he had followed the trotting Tasmanian tiger in 2015.

The usage and expansion of binomial nomenclature in cryptozoology

By Chris Forbes and Raven Heather

Ever since the inclusion of zoological outcasts such as the High-Finned Sperm Whale into *Systema Naturae* (Linnaeus 1768), there have been attempts to apply binomial or "scientific" names to unrecognised species of wildlife. However, it is most common for an animal to receive an official "species description" and accompanying binomial only after a holotype specimen - usually a physical specimen - has been acquired. As it is hard (but not always impossible) for mainstream zoologists to deny the existence of an animal of which a physical specimen is available for study, most cryptids cannot be binomially named the standard way.

However, the use of binomial names can still be incredibly valuable for cryptozoology, for the same reasons it can be valuable for mainstream zoology. A single cryptozoological species can have many different names, coined by the people native to its environment, eyewitnesses, cryptozoologists, and/or the media. This can cause a great amount of unnecessary confusion. Just as binomial nomenclature brought order to the list of officially-recognised animals by assigning each one a single unambiguous label, the same should be encouraged for animals still awaiting such recognition.

Thankfully, there is an existing precedent to legitimize the establishment of a binomial name in the absence of a physical specimen. In Palaeontology, binomials are still applied to animals which have only left indirect proof of their existence in the fossil record, in the form of trackways for example. Since many cryptids *also* leave trackways and other unfossilised "trace fossils", not to mention forms of non-physical evidence unavailable to fossil species (eyewitness accounts, photographs, videos, sonar readings, etc), it can only be an indefensible double-standard to suggest that they do not deserve similar binomial treatment (Heuvelmans 1982). And while granting binomial names to cryptids may remain controversial, it has been suggested that this practice is not necessarily at odds with the rules of the ICZN (Woodley *et al* 2009).

A strong advocate for the application of binomials in cryptozoology was Bernard Heuvelmans, and within his two most infamous works we can find examples of this - along with examples of cryptids still in need of such names.

In *On The Track of Unknown Animals* (Heuvelmans 1958), the first cryptid to receive this treatment is the Yeti:

> "*It is high time that the snowman was also recognised by a scientific name and since the palaeontology of giant primates is so slender, I would give it a new name,* Dinanthropoides nivalis, *or 'terrible anthropoid of the snows.' If one day its teeth are examined and found to be identical with those of the* Gigantopithecus, *its name will have to be changed, according to the rule of priority, to* Gigantopithecus nivalis, *the present species being no doubt quite distinct from the Pleistocene primate from Kwangsi.*"

What's important is that he provides this binomial only after providing an incredibly detailed and in-depth analysis of what the Yeti is. After coining *Dinanthropoides nivalis* he spends the next paragraph listing its characteristics, to an extent that makes it impossible to confuse with any other animal. Just as the purpose of a holotype specimen is primarily to prove that a species is genuinely unique, we should only apply binomials to cryptids which, from the information available, are clearly identifiable as a distinct species.

Heuvelmans is even more clear about this with his second example, the Spotted Lion, on page 443:

> "*Paradoxically enough, the same systematicians who are so ready to create new species on such trivial data, refuse to admit that the spotted lion is a separate variety, even though the differences observed in its size and marking are so distinct that they may well justify the creation of a new species, which could have no better name than* Leo maculatus."

In light of updates to Big Cat taxonomy, this binomial has since been adjusted to *Panthera leo maculatus*, now a lion subspecies rather than a species (Heuvelmans 1986).

Heuvelmans covers many cryptids in *On The Track* that fulfill this criteria of a detailed, distinct description but have not yet been granted a binomial by anyone yet. Going through every example would take us far too long, but some particularly prominent cryptids can be reviewed.

One chapter of Heuvelmans' work details New Zealand's Waitoreke - the country's only native land mammal. This small otter-like or seal-like animal has glossy brown fur and may be a monotreme, and if so, it is the only monotreme outside of Australia and New Guinea. One of the first naturalists to learn of it was Walter Mantell around 1850, and so our suggested binomial for it is *Aotearoatherium mantelli* (Mantell's Aotearoa beast - Aotearoa being the Maori name for New Zealand).

A cryptid which only made a short cameo in *On The Track* but which we now know far more about (and have high hopes for) is the Mapinguari of South America. While Heuvelmans believed this to be a primate, research conducted in more recent years by David Oren indicates that this animal - which stands 2m tall, produces a strong unpleasant smell, has long claws and long coarse fur, and can alternate between a bipedal and quadrupedal stance - is a late-surviving ground sloth, most likely a Megalonychid (Oren 2001). Without a skeleton to analyse it cannot be determined if the Mapinguari belongs to any known fossil genus of Megalonychid, and until such analysis is possible, it should probably be treated as its own separate genus within the Megalonychid family. We can give the Mapinguari the binomial *Nychotherium oreni* (Oren's claw beast).

Moving on to Africa, Heuvelmans introduces a strange and aggressive big cat, distinct from the lion or leopard, known as the Mngwa or Nunda. Heuvelmans' conclusion as to its identity, based on its appearance and its behaviour of purring instead of roaring, is that it's a giant subspecies of the elusive African Golden Cat *Caracal aurata*. We see no issue with this hypothesis, and in reference to the Mngwa's lion-like size and ferocity, we suggest naming this subspecies *Caracal aurata leo*.

One final cryptid shown in *On The Track* which we will coin a much-overdue binomial for is what Heuvelmans calls the 'Congo Dragon', nowadays referred to primarily by one of its local names, the Mokele-mbembe. This semi-aquatic quadrupedal reptile reaches a maximum length of around 9 meters, has a brownish colour, and has a characteristically long neck. Much of what we now know about it has come from Roy Mackal's expeditions to the Congo in the 1980s, although even today it is uncertain if this extraordinary species taxonomically belongs with the Varanid lizards, the Trionychid turtles, or even the Sauropods (Shuker 2016). Thus we will give it a taxonomically "neutral" binomial - *Aquamanyama mackali* (Mackal's lake beast).

A decade after *On The Track of Unknown Animals* came its sea sequel, *In The Wake of the Sea Serpents* (Heuvelmans 1968). Here we again see Heuvelmans apply binomials to unrecognized species, but here he's more methodical about it. After performing a detailed analysis of nearly six hundred alleged sea-serpent sightings, Heuvelmans concluded that they represented up to nine distinct types - seven of which he felt confident in the existence of. As these seven had each been connected to a large number of eyewitness accounts, a detailed description could be provided for all of them. However, only five of these were granted a binomial name. First the two species he classifies as Pinnipeds:

> "*I therefore propose to give the Long-necked sea-serpent (which was known to the ancients as physeter, a word which is now, alas, the generic name for the sperm-whales) the name of* Megalotaria longicollis, *or 'the big sea-lion with a long neck'. The Merhorse I propose to christen* Halshippus olaimagni, *or 'the Sea-horse of Olaus Magnus', because the Norwegian bishop was the first to publish a description of it.*"

And second, the three species he classifies as Archaeocetes:

> "*Each of these Archaecoceti is clearly enough defined to be given a scientific name. The Super-otter, first described by Hans Egede, I propose to call* Hyperhydra egedei, *or 'Egede's Super-otter'; the Many-humped,* Plurigibbosus novaeangliae, *or 'that-with-many-humps of New England', and the Many-finned,* Celioscolopendra aeliani, *or 'Aelian's cetacean centipede', because it is the same animal that the Greek writer described before Rondelet did so.*"

Heuvelmans' hypothesis that each of these five types represents a distinct species may not have aged well. In more recent years it has been suggested that the 'Longneck' and 'Merhorse' are in fact two sexes of the same species, and that the 'Super-Otter' and 'Many-humped' are also synonymous (Coleman *et al* 2003). Since all five binomials were coined at the same time in the same work, it's ambiguous which ones should take priority in these cases. We suggest that the best approach would be to give priority to binomials assigned to types with more sightings - in other words, *Megalotaria longicollis* for the species containing both the Longneck and Merhorse, and *Plurigibbosus novaeangliae* for the species containing both the Many-Humped and the Super-Otter.

Now let's talk about the two types which Heuvelmans felt confident about the existence of, and had found enough sightings to provide a detailed description of, yet did not feel confident in assigning a binomial to.

The first of these is the Marine Saurian, identified from 9 sightings (although one of these has since been exposed as a hoax, and many more sightings have been recorded in the years since then, putting the current total somewhere in the dozens). This species is shaped like a lizard or crocodile, exceeding the size of any lizard or crocodile recognised to exist in the present day. Its grayish-brown or reddish-brown body has two pairs of webbed feet or flippers and it swims in a horizontal (side-to-side) manner like that of a reptile.

On the Marine Saurian's taxonomy, Heuvelmans writes:

> *"The Marine Saurian is probably, as I have said, a surviving thalattosuchian, in other words a true crocodile of an ancient group, a specifically and exclu-sively oceanic one, which flourished from the Jurassic to the Cretaceous Periods. But it could also be a surviving mosasaur, a sea cousin of the moni-tors of today. It would not be surprising if it had survived for so long in the sea, since it is well designed to dive deep and remain unseen."*

It seems this uncertainty about the Marine Saurian's taxonomic position is why Heuvelmans was hesitant to give it a binomial. This is a bit ironic, as the types Heuvelmans *did* name have all since had their taxonomic position questioned in one way or another (Woodley 2008) (Freeman 2019). And since the application of binomials to fossil trackways does not require an identification of the track-maker's taxonomy, we similarly shouldn't need a perfect idea of where a cryptid fits on the tree of life to assign it a binomial - so long as it is known in sufficient detail to clearly determine it to be a unique species.

But, having a rough idea of a cryptid's taxonomy is still useful for coming up with a fitting binomial, so let's ask, to what group of animals *does* the Marine Saurian belong? In the time since Heuvelmans' work, several new hypotheses have emerged, but the most convincing suggestions all still point toward it being a reptile of some sort. Our favorite suggestion is that it may not be a prehistoric survivor as Heuvelmans thought, but rather a member of the Crocodylidae (Marshall 2018). However, we cannot completely rule out the possibility of a lazarus taxa identity, especially considering the presence of Mosasaur remains in post-KPg fossil formations (Gallagher 2005).

One final thing to note is that the Marine Saurian, or at least something quite similar to it, may have been the real-world inspiration for certain versions of dragons such as the Chinese Long (Xu 2018). With all this in mind, our suggested binomial for this species is *Cryptophisaurus long* (Hidden serpent-saurian dragon - while 'saurus' translates to 'lizard', in binomial nomenclature it can refer to any non-avian reptile, even crocodilians such as *Purussaurus*).

The second sea-serpent type Heuvelmans felt confident in but didn't binomially name is the Super-Eel. These are large fish with long, cylindrical bodies and long tapering tails ending in a point. They have large eyes, and a continuous, translucent dorsal fin beginning some distance from the head. Heuvelmans considered 23 of the sightings in his analysis to be of Super-Eels, but as with the Marine Saurian there have been more sightings in the years since.

While the most obvious possibility for the Super-Eel's identity is that it's an enormous species of Eel, over the years there have been many attempts to explain it as being some other sort of fish. Most of these hypotheses do not hold up very well. It is unlikely that the Super-Eel is a new species of Synbranchid, as these fish are not known to exceed 150 cm in length or tolerate saltwater, and oversized Frilled Sharks or Oarfish cannot explain some sightings, such as that made by the entire crew of the ship *Pauline* in 1875. Additionally, Frilled Shark specimens still have yet to exceed a maximum reported length of 196 cm (López-Romero et al 2020), nor is there any strong evidence of Oarfish exceeding 8m (McClain et al 2015). Additionally, the manner in which Oarfish have been observed to swim is distinct from what descriptions of Super-Eel locomotion suggest (Shuker 2003).

Perhaps some of these explanations can apply to the smallest of observed Super -Eels. However, we feel there is still a "true" Super-Eel which must be considered, the species seen by the crew of the *Pauline* and many others. The most fascinating account of this species in the years since Heuvelmans' investigation occurred at a depth of 220 meters (Shuker 2023), supporting his suspicions that it lives below 100 fathoms - in other words that, like the Giant Squid, it's native to the ocean's mesopelagic zone.

Large and elongated body forms have independently evolved many, *many* times among fish, so we cannot say we are 100% certain about what the Super-Eel is. However, by process of elimination, we think that it is most likely a gigantic member of the Anguilliformes. Pinpointing which family of Eels is responsible is more difficult - both Congers and Morays already have recognised species exceeding 3m in maximum length, making either family a good candidate. Before Heuvelmans' great sea-serpent investigation, deep-sea giant eels were promoted as an identity for sea-serpents by ichthyologist Anton Bruun, and so we shall name this species *Titanguilla bruuni* (Bruun's titan eel).

A second cryptozoological species also appears to be present in this category. Heuvelmans considered it plausible that some Super-Eel sightings could be caused by an unrecognised species of elongated shark. This cryptid, which he nicknamed the "Snark", was likely the animal caught in the Gulf of Maine and described in exceptional detail by Captain S. W. Hanna in 1880. Hanna's description of the immense fish leaves little to the imagination, and sets it apart not just from the non-cryptozoological candidates mentioned above, but also from the "true" Super-Eel. Setting it apart from descriptions of the latter are features such as shark-like skin, a triangular dorsal fin, and shark-like gill slits. As Heuvelmans concluded, these features suggest that the specimen belonged to a species of elongated shark, perhaps related to the aforementioned Frilled Shark although not similar enough to belong in the same genus. Our suggested binomial for this species is *Ophiselachus hannai* (Hanna's snake-shark).

Sea serpents are not the only elusive marine monsters to appear in *In The Wake of the sea serpents*. One rather striking cameo appears in the book's chapter on mystery carcasses:

> *"It is disturbing to find this tale of hairy sea-monsters bigger than whales coming so confidently from places so far apart. Of course they might sometimes be based on the carcass of a sea-serpent with real hair - that is to say a large un-known mammal - being washed up. But such monsters can be explained without this theory, for we have physical proof that there were once similar creatures in the sea ... some specialists have calculated that Carcharodon megalodon must have been between 80 and 120 feet long. This figure is based on the relative size of the teeth of various related species, but in fact for mechanical reasons giant species are almost always proportionately longer than the normal, so the larger figure is most likely to be right. It may well even be too small. The story of the coelacanth shows that there is no a priori reason why such enormous monsters should not survive. After all, we should not have*

known that there still are savage and voracious 60-foot sharks if their teeth had not been discovered by chance. The past existence of even larger Carcharodons *shows that sharks may be bigger than the biggest whales, in which case may not the legend of hairy sea-monsters be based, as it was in Flacourt's story, on decomposed carcases, and may they not be the carcases of giant selachians?"*

Some parts of Heuvelmans' statement have not aged well. We now know that Megalodon, which is now placed in the genus *Otodus* rather than *Carcharias*, reached a maximum known length of only around 20.3m, at most 24.2m (Perez et al 2021). Heuvelmans' belief that the Great White can reach 60 feet was also erroneous, as he had assumed a pair of Megalodon teeth collected by the HMS *Challenger* to have come from a Great White instead - a clever but incorrect explanation for the controversial age of the teeth.

Nonetheless, evidence does exist of giant cryptozoological sharks in modern times. Some of this evidence, in particular giant bite marks, could point toward Great Whites growing beyond their "official" maximum size, or a descendant of Megalodon's smaller relative *Otodus chubutensis* (Hawthorne 2021). And then of course there is the remote possibility, especially when considering eyewitness observations like that which occurred in 1918, that this giant shark is a descendant of *Otodus megalodon* itself.

If *Otodus chubutensis* or *Otodus megalodon* has survived into modern times, it must have evolved to become quite different from the fossil-record forms or else its presence would be much easier to confirm. In particular it would have to adapt to live in deeper waters, perhaps being forced to do so due to ecological competition with Great Whites, Orcas and non-abyssal Sea Serpents. In other words, a modern "Megalodon" would have to be considered a new species in the *Otodus* genus.

One could argue that a binomial for this cryptid has already been coined over two centuries ago: *Halsydrus pontoppidani*, given by Patrick Neill in 1809 to the 16.8m shark carcass discovered on the Scottish island of Stronsay in late September of the year prior. The largest species of shark known in this region is the Basking Shark, which does not grow much more than 12m long. If it is ever confirmed that the Stronsay carcass is a specimen of a late-surviving *Otodus*, then by law of taxonomic priority, the correct name for the species will be *Otodus pontoppidani*. If, on the other hand, the Stronsay carcass was simply the largest Basking Shark or most lost Whale Shark ever and a late-surviving *Otodus* exists separately, we suggest giving the latter the simple binomial *Otodus modernicus*.

Moving on from Heuvelmans' research we can look to another highly influential cryptozoological work to find our final past example of a binomial name being applied to a cryptid. Released in 1995, Karl Shuker's *In Search of Prehistoric Survivors* was one of the most superb and comprehensive works since those of Heuvelmans explained above. While its contents have since been improved even further in both quantity and quality in its modern, updated version *Still In Search of Prehistoric Survivors*, there is one particular analysis which first appeared in the original which is relevant to this question of binomial naming.

Native to Central Africa, the Makalala is - or rather was, as its almost certainly extinct now - the tallest bird to live in modern times, exceeding the height of the Ostrich *Struthio camelus*. What makes this cryptid stand out, in addition to its detailed description, is that within its habitat range, native chiefs used Makalala skulls as helmets. This raises the possibility that, even if the animal itself is long gone, proof of its existence could still be acquired if someone were to search for one of these helmets. In *In Search*, Shuker used the Makalala's characteristics to pinpoint its taxonomic position, and so coined for it a binomial:

> *"Could the makalala, therefore, be some form of extra-large secretary bird, not necessarily as tall as the Wasequas state (their fear of it would certainly inflate their estimate of its height), but much bigger than today's single known species? If so, a suitable scientific name for it would be* Megasagittarius clamosus - *'the noisy, giant secretary bird'."*

It is unfortunate that we only see this one example of a binomial name being coined, in a work covering such a vast array of mystery animals. Let's talk about some of the ones Shuker covered, which are still in need of such a name.

One of the biggest breakthroughs in Central African cryptozoology to be clarified in the years between *On The Track* and *In Search* is the distinction of the "Mokele-mbembe" into two unrelated but commonly-confused animals, the hornless, long-necked "true" Mokele-mbembe, and the horned, short-necked Emela-ntouka. While Heuvelmans' conclusions about the Mokele-mbembe's overall appearance were mostly correct, it seems that some parts of his original description, in particular the horn on its head, were likely the result of these two cryptids being muddled up with one another. While it was once thought that the Emela-ntouka was a Ceratopsian dinosaur, in more recent years new evidence is pointing toward it being a mammal. In particular, it seems likely to be a member of the Rhinocerotidae (and unlike both of Africa's officially-recognised Rhinoceros species, which possess two horns, the Emela-ntouka seems to have only one). A taxonomic position within the Arsinoitheriidae is also possible, although this seems less likely.

According to local testimony, the Emela-ntouka is known to use its deadly horn to kill elephants and buffaloes. Indeed this is where the name "Emela-ntouka", meaning "killer of elephants", originates. In reference to this - and its historical confusion with Central Africa's other immense semi-aquatic cryptozoological amniote, the Mokele-mbembe - we suggest giving it the binomial *Ceratophoneus mokelemimus* ("Mokele-mimicking horned assassin").

While the Global North has a relative scarcity of cryptids that are both large, obscure, and promising, it is not devoid of them. One of the most intriguing mystery animals within the bounds of the Contiguous United States is a giant salamander that has been seen - and on some occasions even caught - in the Trinity Alps region of California. This Salamander seems similar to the Giant Salamanders of the genus *Andrias*, and while the only officially-recognised members of this genus in the present day are restricted to East Asia, the fossil record provides an obvious candidate for this cryptid's ancestor. During the Oligocene and Miocene, North America was home to *Andrias matthewi*, a Salamander which reached up to 2.3 meters in length (Naylor 1981). Since no fossils of *A. matthewi* have yet been found within the state of California, we suggest naming the Trinity Alps Salamander *Andrias californiensis*.

In the present time, two living species of Coelacanth are recognised to exist. These are *Latimeria chalumnae*, native to the West Indian Ocean off the coast of Africa, and *Latimeria menadoensis*, native to Indonesia. Both are former cryptids, which were known about by native peoples long before their "official" discovery. So it shouldn't come as a surprise that evidence exists of a third living Coelacanth species. Unlike the prior two, this Coelacanth appears to be native to the waters of the New World. Scales and silver ornaments indicate that, besides this unexpected location, the fish is otherwise quite similar to the two recognised Coelacanth species. Taking everything into consideration, we think this is most likely a distinct species within the genus *Latimeria*, and thus suggest the name *Latimeria occidentalis* (Western *Latimeria*).

The Malayan Tapir *Tapirus indicus* is the only species of tapir native to Asia, and two subspecies are officially recognised to exist within an "official" habitat range encompassing the Malay Peninsula and Sumatra. However, there have been a number of unconfirmed reports suggesting that a third living subspecies may exist on the island of Borneo, which has "officially" been tapir-free for thousands of years. The existence of this subspecies may have even come very close to confirmation half a century ago. In 1975, Indonesia's Antara News Agency reported the capture in Kalimantan of what sounds at first glance to be a totally bizarre animal. Its supposed combination of tiger, lion, bird and goat features earned it the name 'tigelboat', however, a methodical analysis of its description reveals it is quite similar to a normal Malayan Tapir. It seems that the description of this 'tigelboat' had simply come from someone who was totally unfamiliar with the animal. However, there is one notable difference between the recognised subspecies of *T. indicus* and the tigelboat. The latter was said to be maned and bearded. As Shuker explains:

> "*It is true, of course, that, whereas its three New World relatives are maned, the Malayan tapir normally lacks any notable extent of hair upon its neck (though juveniles are somewhat hairier than adults); equally, it is not bearded. However, during its existence upon the island of Borneo for 10,000 years since the end of the Pleistocene, totally separated from all other populations of* T. indicus *elsewhere in Asia, it is probable that a Bornean contingent of Malayan tapirs would evolve one or two morphological idiosyncrasies (just as isolated populations of many other widely distributed animal species have done). Nothing very spectacular, but enough to permit differentiation from all other* T. indicus *specimens; such features could readily include a mane or a beard or both.*"

In other words, Borneo's tapir seems to be a distinct subspecies, for which we suggest the simple binomial name *Tapirus indicus tigelboat*, referencing the specimen which unfortunately failed at the time to attract any scientific interest and eventually disappeared.

Whilst some may disregard binomial naming in cryptozoology (and even the subject itself), for the aforementioned reasons stated in the introduction, this piece looks to provide an ulterior angle to this area of research. For cryptozoology more so than its mainstream counterpart, communication is at the core of the study. This key to success can be aided in much the same way as binomial nomenclature has aided mainstream biology for centuries, it makes for an inter-language understanding of nature, and a greater level of communication which cannot be discarded. Cryptozoology works in part with many indigenous communities most of which have very different naming and systems of classifying and grouping animals as the wider scientific world. In order for this area to be fully explored, and too with the rest of the world, communication on this level is crucial.

Sources cited:

- **Coleman L, Huyghe P. 2003.** The field guide to lake monsters, sea serpents and other mystery denizens of the deep. New York (NY): Tarcher.
- **Freeman R. 2019.** Adventures in Cryptozoology: Hunting for Yetis, Mongolian Deathworms, and Other Not-So-Mythical Monsters: Volume I. Coral Gables (FL): Mango Publishing Group.
- **Gallagher WB. 2005.** Recent mosasaur discoveries from New Jersey and Delaware, USA: stratigraphy, taphonomy and implications for mosasaur extinction. Netherlands Journal of Geosciences. 84:241–245.
- **Hawthorne M. 2021.** Monsters & Marine Mysteries. Las Vegas (NV): Far From The Tree Press.
- **Heuvelmans, B. 1958.** On the Track of Unknown Animals. London (UK): Rupert Hart-Davis.
- **Heuvelmans B. 1968.** In the Wake of the Sea-Serpents. New York (NY): Hill and Wang.
- **Heuvelmans B. 1982.** What is cryptozoology? Cryptozoology. 1:1–12.
- **Heuvelmans B. 1986.** Annotated checklist of apparently unknown animals with which cryptozoology is concerned. Cryptozoology. 5:1–26.
- **Linnaeus C. 1758.** *Systema Naturae per regna tria naturae, secundum classes, ordines, genera, species, cum characteribus, differentiis, synonymis, locis. Editio decima, reformata.* Stockholm: Holmiae : Impensis Direct. Laurentii Salvii.
- **López-Romero FA, Klimpnger C, Tanaka S, Kriwet J. 2020.** Growth trajectories of prenatal embryos of the deep-sea shark *Chlamydoselachus anguineus* (Chondrichthyes). J. Fish. Biol. 97:212–224.
- **Marshall C. 2018.** 21st Century Sea Serpents. Animals & Men. 64/5:37–69.
- **McClain CR, Balk MA, Benfield MC, Branch TA, Chen C, Cosgrove J, Dove AD, Gaskins LC, Helm RR, Hochberg FG, Lee FB. 2015.** Sizing ocean giants: patterns of intraspecific size variation in marine megafauna. PeerJ. 3:p.e715.
- **Naylor BG. 1981.** Cryptobranchid Salamanders from the Paleocene and Miocene of Saskatchewan. Copeia. 1981(1): 76–86.
- **Oren DC. 2001.** Does the Endangered Xenarthran Fauna of Amazonia include remnant ground sloths? Edentata. 4:2–5.
- **Perez VJ, Leder RM, Badaut T. 2021.** Body length estimation of Neogene macrophagous lamniform sharks (*Carcharodon* and *Otodus*) derived from associated fossil dentitions. Palaeontologia Electronica,

24:a09.

- **Shuker KPN. 1995.** In search of prehistoric survivors. London (UK): Blandford.
- **Shuker KPN. 2003.** The beasts that hide from man. New York (NY): Paraview Press.
- **Shuker KPN. 2016.** Still in search of prehistoric survivors. Greenville (Ohio): Coachwhip Publications.
- **Shuker KPN. 2023.** An 'Ell of an Eel. Fortean Times. 427:21.
- **Woodley MA. 2008.** In the wake of Bernard Heuvelmans: an introduction to the history and future of sea serpent classification. Bideford (UK): Centre for Fortean Zoology Press.
- **Woodley MA, Naish D, Shanahan HP. 2009.** How many extant pinniped species remain to be described? Historical Biology. 20(4):225–235.
- **Xu DC. 2018.** Mystery Creatures of China: The Complete Cryptozoological Guide. Greenville (Ohio): Coachwhip Publications.

If it's there, could it be a bear?
Floe Foxon

ABSTRACT
It has been suggested that the American black bear (*Ursus americanus*) may be responsible for a significant number of purported sightings, of an alleged unknown species of hominid, in North America. Previous analyses have identified correlation between 'sasquatch' or 'bigfoot' sightings and black bear populations in the Pacific Northwest, using ecological niche models and simple models of expected animal sightings. The present study expands the analysis to the entire US and Canada, by regressing sasquatch sightings on bear populations in each state/province, while adjusting for human population and forest area in a generalized linear model. Sasquatch sightings were statistically significantly associated with bear populations such that, on the average, every 1,000 bear increase in the bear population is associated with a 4% (95% CI: 1%–7%) increase in sasquatch sightings. Thus, as black bear populations increase, sasquatch sightings are expected to increase also. On the average, across all states and provinces in 2006, after controlling for human population and forest area, there were approximately 5,000 bears per sasquatch sighting. Based on statistical considerations, it is likely that many supposed sasquatch are really misidentified known forms. If bigfoot is there, it may be a bear.

INTRODUCTION
The United States and Canada feature nearly 20 million km^2 of land, hosting hundreds of mammal species in its woodlands, prairies, boreal forests, and along its coasts (**Kays & Wilson 2009**). Due to this size and diversity, it is unlikely that the North American faunal catalogue is complete, with numerous insects and possibly small mammals remaining to be discovered. While most scientists believe that the existence of a large hominid species in North America, as-yet unrecognised in conventional science, is not likely (**King & Greenwell 1983**), some authors have entertained this possibility in the field of research called 'hominology'.

Hominology, as formulated by Boris Porshnev and Dimitri Bayanov, alleges that hairy, bipedal primates not recognised in conventional zoology, *do* exist and are relicts of *Homo neanderthalensis*, banished to the wild through competition with *H.*

sapiens (**Bayanov 2012**). One such mystery hominid reported in North America, is variously dubbed 'sasquatch' after West Coast First Nations tradition (**Heuvelmans 1986**), 'bigfoot' to Westerners, and the *nomen dubium Gigantopithecus canadensis* to cryptozoologists, who believe that rather than a hominin, this animal is a relict Pleistocene pongid (**Heaney 1990**). Purported sasquatch sightings number in the thousands (**Bigfoot Field Researchers Organization 2023**), making bigfoot a prominent anthrozoological phenomenon.

Hominologists describe six lines of evidence to support the above claims (**Bayanov 2012**): (1) descriptions of wild men in the natural history texts of ancient Roman and Arab philosophers and medieval Europeans; (2) folklore and mythology; (3) ancient and medieval art; (4) footprints, tracks, etc.; (5) photographs; and (6) eyewitness testimony. These are all categories of indirect evidence (i.e., testimonial and circumstantial).

Concerning (1), while some legendary animals in ancient texts were likely inspired by encounters with real forms (e.g., the Greco-Roman griffin and its probable inspiration in fossil ceratopsians (**Mayor 1991**)), the existence of other beings catalogued in ancient and medieval bestiaries have not been treated with such conviction. Similarly concerning (2), while various cross-culture traditions, myths, and legends have been identified in association with mystery hominids including sasquatch (**Bayanov 2011**), many characters in folklore such as magical witches and sorcerers do not correspond to counterparts in reality (**Colarusso 1983**). In hominology, folklore is interpreted as the emotional expression of lived experiences (**Bayanov & Bourtsev 1976**), therefore sasquatch traditions are seen as evidence for sasquatch existence. However, many of these folkloric references are extremely heterogeneous in their descriptions of mystery hominids.

Concerning (3), 'hairy man' petroglyphs in the Tule River Tribe Reservation, approximately 1,000 years old, have been interpreted as pictorial representations of bigfoot (**Strain 2012**). However, other Native American petroglyphs depicting animals such as humanoid frogs have not received such literal interpretations, which may indicate a kind of confirmation bias in hominology research.

Concerning (4), numerous casts and measurements of tracks and footprints attributed to sasquatch have been presented (**Napier 1976**, 'Tables'), some apparently featuring primate-like dermal ridge patterns, sweat pores, and sole pads (**Krantz 1983**; **Cachel 1985**). These have been criticised as hoaxes constructed with modelling clay and the 'latex-and-kerosene expansion method' of preserving details of [human] footprints while greatly increasing their size (**Baird 1989**; **Bodley 1988**). It has been suggested that features such as 'sweat pores' are casting artefacts such as air bubbles (**Freeland & Rowe 1989**). Genetic and microscopic analyses of supposed hairs, faeces, and other specimens attributed to sasquatch have been variously identified as synthetic fiber (**Winn 1991**; **Somer 1989**), or material from known forms such as cervids, bovines, and ursids (**Federal Bureau of Investigation 2019**; **Bryant & Trevor-Deutsch 1980**; **Coltman & Davis 2005**; **Sykes et al. 2014**; **Hart 2016b**,a).

Concerning (5), the 'Patterson-Gimlin film' is perhaps the 'best' photographic evidence for sasquatch. This notorious 16mm motion picture purportedly depicts an unknown hominid over six feet tall in California (**Munns 2014**; **Kelsey 2022**; **Discovery 2022**). The film apparently was not spliced or edited (**Munns & Meldrum 2013**), but many have noted the imposing likelihood that the film subject is a suited actor.

This leaves (6), the testimony of thousands of eyewitnesses. The aim of the present study is to better understand what thousands of people see when they report a sasquatch encounter. Besides hoaxes, the leading hypothesis among sceptics is that when eyewitnesses report seeing bigfoot, they are actually seeing bears (**Nickell 2013**). To understand how this can be the case, it is necessary in the following to describe the characteristics of bears in North America, and how these are similar to purported characteristics of sasquatch.

North America features three bear species, chiefly among them being the American black bear (*Ursus americanus*). The black bear is a large tetrapod (lengths up to 6 feet or more and masses up to 300kg), with a pelage of various shades of black, brown, red, blue, blond, and white (**Burton 1998**). Morphologically, the black bear is stocky with strong musculature around its thick neck, shoulders, and legs (**Clark et al. 2021**). The tail is short and therefore may be difficult to notice.

The geographic distribution of the black bear is widespread, ranging all across Canada and Alaska, the East and West coasts of the United States, south into Mexico, and with various isolated populations in-between (**Beer & Morris 2004**). Although the present distribution is much diminished from the historical range, the black bear has experienced a notable expansion since the 1990s (**Scheick & Mccown 2014**). Black bears may inhabit scrub, swamps, and mountains, but are primarily found in forest and woodland (**Beer & Morris 2004**; **Burton 1998**).

Black bears have opportunistic foraging habits; being omnivores, they consume various plant material (including berries, fruits, nuts, and grasses), other mammals (neonate ungulates), fish, invertebrates, and carrion (**Beer & Morris 2004**). This opportunism means habituated and food-conditioned bears can, and will, approach people and buildings in search of anthropogenic food sources such as agriculture and garbage (**Herrero 2018**, Chapter 4).

Although this dietary range adaptation has allowed black bears to remain widespread (**Clark et al. 2021**), it is also the primary cause of confrontation between humans and bears in areas such as national parks (**Herrero 2018**, Chapter 4). Black bears are easily food conditioned (**Brown 1993**, Chapter 4), and because of this they are considered pests in much of North America (**Burton 1998**).

The presence of the black bear is indicated by such signs as overturned logs and broken twigs and branches (**Burton 1998**), overturned surface stones, tracks, hairs, and claw and tooth marks left on trees (**Herrero 2018**, Chapter 4). These bears are solitary and active at night, and may be heard vocalising through grunts, growls, woofs, and howls (**Beer & Morris 2004**), as well as snorts, roars, coughs, and squeaks, all indicative of emotion (**Brown 1993**, Chapter 4). Black bears also have a pronounced odour (**Brown 1993**, Chapter 3). They are inquisitive (**Brown 1993**, Chapter 4) and intelligent animals (**Vonk et al. 2012**), capable of forming natural concepts at concrete, intermediate and abstract levels. Bears in general are plantigrade and are known to stand upright and walk on their hind legs; some bear species are known to ambulate over 400 meters bipedally (**Brown 1993**, Chapter 4). Black bears are also strong swimmers, outstanding climbers, and can run at speeds of up to 13 meters per second (**Brown 1993**, Chapter 4).

It is precisely the above characteristics of the black bear that make this animal a likely candidate for many purported sasquatch sightings (**Nickell 2013**). Indeed, **Brown (1993**, Chapter 5) lists numerous lines of evidence connecting sasquatch traditions with bears, including reported size, strength, and speed; bipedal locomotion; leaving human- or ape-like footprint outlines; tracks found and encounters occurring in bear habitats; and having the same colouration.

Cryptozoologist Ivan **Sanderson (1967**, Chapter 18) also described sasquatch as "forest dwellers", sightings of which fall within the distribution of the Earth's vegetational belts (i.e., where bears and bear foods are found). Other reported characteristics of sasquatch include distinctive smell and various vocalisations; eyes "like a bears"; "reddish", "dark brown", "light brown", and other fur colours; consumption of berries and other vegetation; rolling boulders; and raiding of houses for food (**Sanderson 1967**, Chapters 2–4), are all consistent with black bears in North America, as described above. Of course, an alternative interpretation of these correlates is that the sasquatch is an animal distinct from the black bear but is similar to it in many ways, including habitat and the characteristics described previously.

Existing studies have explored the possible link between the American black bear and bigfoot. **Blight (2005)** used probabilistic models to examine the relationship between sasquatch sightings and black bear populations in the Pacific Northwest (PNW; including Alaska, Montana, Oregon, Washington, Northern California, and Idaho) and the rest of the US. The method used was a simple calculation of the bivariate correlation coefficient between sasquatch sighting frequency and black bear population density. That study identified positive correlation between black bear populations and sasquatch sightings. One criticism of this study is that the test used assumed a Gaussian distribution for count data. Furthermore, only the US was considered, despite sasquatch sightings being considerably more widespread across the entire US and Canada.

In another study, **Lozier et al. (2009)** identified a high degree of overlap in predicted geographic distributions for black bears and sasquatch. The method used was to compare results from ecological niche models (ENMs) for each of sasquatch and black bears in the PNW. ENMs are tools used in conservation to predict geographic ranges of animals; they take as input georeferenced data (i.e., events with geographic coordinates) and environmental data layers constructed from bioclimatic variables (e.g. annual mean temperature), and output a predicted geographic distribution for the animal (**Sillero et al. 2021**). Although the purpose of the study by **Lozier et al. (2009)** was to highlight potential flaws in these models when errors are introduced by inaccurate specimen identities and georeferencing, the analysis is supportive of a statistical association between sasquatch and black bears; the authors state explicitly that the predicted sasquatch distribution may be seriously biased if many (or all) of the sightings represent misidentified black bears, hence the overlap. The **Lozier et al. (2009)** analysis considered only data from the PNW (Washington, Oregon and California).

The present study builds upon the works of **Blight (2005)** and **Lozier et al. (2009)** by using more recent data, by expanding the analysis to the entire US and Canada, and more appropriately assuming a count data distribution, as opposed to a Gaussian distribution. Statistical methods (regression modelling) are used to test the hypothesis that sasquatch sightings and populations of black bears are associated, while controlling for potential confounding by human population and forest area.

METHODS

The number of black bears in each US state and Canadian province were sourced from **Table 1.** of **Spencer et al. (2007)**. To the author's knowledge, this represents the latest peer-reviewed, publicly available collection of black bear population data, providing population estimates in each state/province for the year 2006. Because Delaware, Hawaii, Illinois, Indiana, Iowa, Kansas, Nebraska, North Dakota, and South Dakota have no known breeding populations of black bear, the black bear populations were assumed to be zero in these states. Other species of bear were not considered, as their impact on the result is likely to be negligible; black bears are by far the most abundant species, numbering over two times that of all other bear species combined (**Garshelis et al. 2021**).

Table 1.
Regression Model Parameters (* $p < 0.05$, ** $p < 0.0001$)

Variable	Regression Coefficient ± Standard Error
Intercept	0.5 ± 0.2 *
Black Bear Population	$(4 \pm 2) \times 10^{-5}$ *
Human Population	$(1.2 \pm 0.2) \times 10^{-7}$ **
Forest Area (km^2)	$(-7 \pm 3) \times 10^{-6}$ *

The number of sasquatch sighting reports in each US state and Canadian province were sourced from the Bigfoot Field Researchers Organization's Geographic Database of Bigfoot/Sasquatch Sightings & Reports (2023). These data consist of eyewitness testimonials, mostly from the second half of the 20th century to the present. To maximise the validity of the analysis, only sightings in the year 2006 were considered such that bear populations and sasquatch sightings were date-matched in these analyses.

Human population statistics for each US state and Canadian province in the year 2006 were obtained from the **United States Census Bureau (2021)** and **Statistics Canada (2022)**, respectively. Forest area estimates (in km^2) for each US state in 2006 were sourced from the **USDA Forest Service (2006)**, and estimates for Canadian provinces were sourced from World Atlas (**Sawe 2017**). For choropleth map plotting, a geojson map of the US and Canada was sourced from **Cartograhy Vectors (2022)**.

Four US states (Rhode Island, Texas, Wisconsin, and Wyoming) and four Canadian provinces (Alberta, Newfoundland and Labrador, Northwest Territories, and Nova Scotia) were missing data on bear population. These were necessarily excluded from analyses.

To test the hypothesis that sasquatch sightings and populations of black bears are associated, a regression model was implemented. This model regressed the number of sasquatch sighting reports in each state/province in 2006 (the dependent/response variable) on the black bear population in each state/province in 2006 (the independent/predictor variable). An intercept term was included to model the possibility of sasquatch sightings in states with no bears (i.e., to model the value of the response variable when the predictor equals zero).

Two more variables were included in the model to reduce confounding and increase the internal validity. These variables were the human population and forest area in each state/province. The more people in a given state/province, the more likely a sighting is to occur. To account for this, the human population in each state/province in 2006 was included as an independent variable in the model. Similarly, the area of forest land in each state/province is likely to impact the number of sasquatch sightings due to the relationship between area and population density, and because black bears are primarily found in forest and woodland (**Beer & Morris 2004; Burton 1998**).

Consequently, the total amount of forest area (in km^2) in each state/province was also included as an independent variable in the model.

Because these data (number of sightings) are count data (i.e., countable quantities), it was necessary to use a model appropriate for count data. The model used was a generalized linear model assuming a Negative Binomial distribution and the log link function, which are appropriate for count data. Variations on this model design (e.g., instead assuming Gaussian and Poisson distributions; including interaction terms and random effects; and excluding the intercept term) were investigated in previous exploratory analyses (**Foxon 2023b,c**). In those exploratory analyses, model fits were assessed by the root mean square error (RMSE; lower is better) and log-likelihood (higher is better). Gaussian models provided the lowest RMSE but it is generally known that Guassian approximations do not correctly describe count data (**Lass et al. 2021**). Interaction terms were not statistically significant and/or provided poorer model fits to the data, and so were not included in the final model. The inclusion of random intercepts for each state/province overfit the model because only one timepoint is used (i.e., just 2006). The final model in the present study (Negative Binomial, log link, and intercept) provided the highest log-likelihood and is arguably the most appropriate model for these data for the reasons described above.

Indeed, statistical models using negative binomial regression techniques describing the abundance of various animals species have been successfully developed in previous studies in the context of conservation biology, with high predictive power for some species (**Pearce & Ferrier 2001**; **Pradhan & Leung 2006**; **Acevedo et al. 2014**).

All analyses were performed in Python 3.8.16 with the packages Numpy 1.21.5, Pandas 1.5.2, Scipy 1.7.3, Statsmodels 0.13.5, and Plotly 5.9.0. All code and data are available in the online Supplementary Information (**Foxon 2023d**). Statsmodels uses the iteratively reweighted least squares (IRLS) method for generalized linear models to reduce the impact of outliers.

This work uses only publicly available secondary data on animal subjects. It did not involve primary (*in vivo*) animal research and so is exempt from ARRIVE guidelines.

RESULTS
Figure 1 shows choropleth maps for the number of sasquatch sightings, black bear populations, human populations, and forest area in the United States and Canada in 2006. At first glance, there are locations such as Florida with very many sasquatch sightings but relatively few bears (at least compared to some other states/provinces). This appears to be inconsistent with the hypothesis that sasquatch sightings and populations of black bears are associated. However, this simple comparison of bears to bigfoot is confounded by the human population and forest area in each state/province. When considering the three maps of bears, people, and forest area simultaneously, the possible associations become more apparent. For example, Florida has relatively few bears and relatively little forest area, but also a relatively large number of people to make sightings. It may be that the combination of fewer bears and less forest area but more people explains the higher number of sightings. Thus, the only way to investigate the possible association between sasquatch sightings and populations of black bears is by considering the three predictors (bears, people, and forest area) simultaneously, which is the purpose of the model.

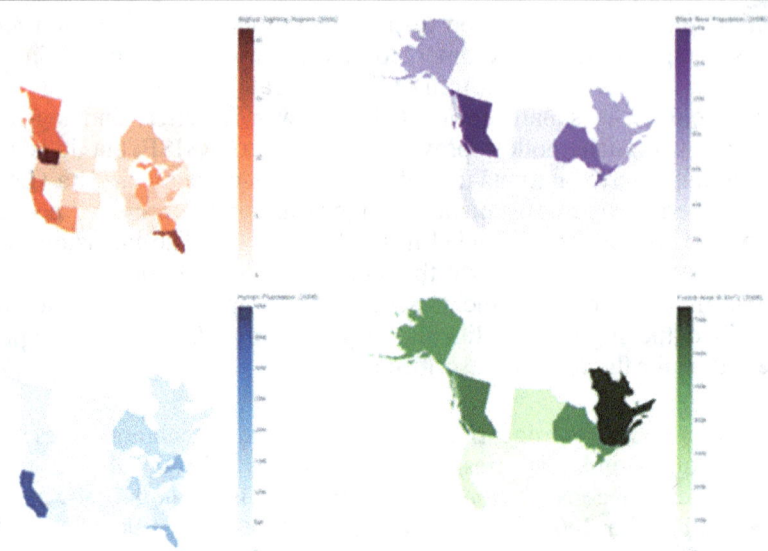

Figure 1.
Choropleth maps for sasquatch reports, black bear (*Ursus americanus*) populations, human populations, and forest area across the United States and Canada.

The results of the model regressing bears, people, and forest area on sasquatch sightings in 2006 are described in **Table 1**. In this model, black bear population was associated with sasquatch sightings such that, on the average, every 1,000 bear increase in the bear population is associated with a 4% (95% CI: 1%–7%) increase in sasquatch sightings.

Thus, as black bear populations increase, sasquatch sightings are expected to increase also. On the average, across all states and provinces in 2006, after controlling for human population and forest area, there were approximately 5,000 bears per sasquatch sighting (note that being an average across states/provinces, this estimate is not accurate for each individual state/province). The RMSE of the model (19) was high relative to the number of bigfoot sightings in 2006 in each state/province (range: 0–21).

DISCUSSION
In the present study, a model was used to investigate the hypothesis that sasquatch sightings are associated with black bear (*Ursus americanus*) populations across the US and Canada. From this model, a positive association was identified between sasquatch sightings and black bears such that, after adjusting for human population and forest area, one sasquatch sighting per year is expected for every 5,000 bears in North America.

While correlation does not equal causation, perhaps the most parsimonious interpretation of these findings is that many supposed sasquatch sightings in North

America may be explained as misidentified black bears. This is logical, because as noted by **Nickell (2013)** and **Brown (1993**, Chapter 5), bears and sasquatch share many characteristics in habitat, appearance (e.g., size, hair/fur coverage, and colouration), attributes (e.g., speed and strength), and behaviour (e.g., bipedal locomotion and curiosity). Nickel also notes that poor viewing conditions, non-expert observation, and expectant attention (i.e. 'seeing' what one anticipates) could explain why some people might confuse bears for mystery hominids.

Of course, hominologists may point to the possibility that sasquatch are animals distinct from bears but which live in the same places and look and act similarly. However, in the absence of autoptical evidence (i.e., not testimonial or circumstantial), there is not at present strong scientific support for the existence of sasquatch as a mystery hominid.

Although preliminary, these findings suggest that sasquatch sightings may have some utility in bear conservation efforts as a form of citizen science. Data from amateurs/non-professionals are already in use in conservation biology/ecology with great success while also being low in cost (**Sumner et al. 2019**; **MacPhail & Colla 2020**; **McKinley et al. 2017**). It may be that collaboration between bear conservationists and bigfoot field researchers could lead to more recent and accurate estimates of bear population and geographic distribution across North America, improving occupancy estimation by reducing non-detection error (false negatives).

Indeed, cryptozoological anecdotes, in plurality, may well contain legitimate zoological insights (**Opit 2017**). **Paxton (2009)** has found that in the context of eyewitness reports of unidentified, large marine animals, cryptozoological anecdotes are amenable to scientific analysis and can be considered data. However, caution must be taken; false-positive (species misidentification) observations can cause overestimation of the distribution or abundance of a species (**Costa et al. 2015**), as in the case of the Eurasian lynx in the Alps (**Molinari-Jobin et al. 2012**), the white marlin in the western North Atlantic (**Beerkircher et al. 2009**), and perching birds in South America (**Gorleri & Areta 2022**; **Gorleri et al. 2023**). The solution may be to utilise more advanced models that simultaneously account for both false negative and false positive identification (**Miller et al. 2011**).

The primary limitation of this study is the potential for residual confounding by factors other than bear population, human population, and forest area. For example, people living, hunting, or hiking in forest areas may also be misidentified as sasquatch. Furthermore, it is important to note that sasquatch sightings have been reported in states with no known breeding black bear populations. Indeed, the intercept term of the model was positive, implying that there are sasquatch sightings even when the number of bears equals zero. Although this may be interpreted as evidence for the possible existence of an unknown hominid in North America, it may also explained by misidentification of other animals (including humans), among other possibilities. Still, the present study does not prove that all sasquatch sightings are bear sightings (nor was this study designed for that purpose). Due to these and other limitations specified in the data sources, the findings of the present study must be interpreted as only approximate and not exact (Indeed, as noted above the RMSE was high). Further research and more data are necessary to improve these estimates.

Strengths of this study include the use of a quantitative model that controlled for possible confounding by human population size and forest area in each US state and Canadian province. Furthermore, positive associations between bear populations and bigfoot sightings were also identified in previous exploratory analyses with variations on the model design and data (**Foxon 2023b,c**), which may suggest that these findings are robust. These findings are also in agreement with the results of previous studies by **Blight (2005)** and **Lozier et al. (2009)**, which similarly identified positive associations between bears and bigfoot using different methods. Therefore, the association is likely to be real. **Blight (2005)** reported only weak positive correlation between bears and bigfoot, whereas the association in the present study is rather more strong. This may be because **Blight (2005)** assumed a continuous probability distribution (as opposed to a count distribution), did not consider Canada, did not date-match the data, and did not control for forest area. The findings of the present study were closer to those of **Lozier et al. (2009)**, who reported a high ecological niche model overlap statistic (i.e., a strong association between bear and sasquatch geographic distributions). Another strength is that in the present study, the analyses considered the entire US and Canada, whereas the previously-published works only considered the US or Pacific Northwest. Furthermore, the data in the present study were date-matched (in 2006) to provide more accurate estimates of the association.

In conclusion, if bigfoot is there, it may be a bear.

Fundings
This work was not supported by any specific grant from funding agencies in the public, commercial, or not-for-profit sectors.

Conflict of intertest disclosure
The author declares that they have no financial conflicts of interest in relation to the content of the article.

Data, script, code, and supplementary information availability
Data and code are available online: https://doi.org/10.17605/OSF.IO/AV3G2

Acknowledgements
The author thanks Julie Sheldon and Rahul Raveendran for helpful comments and suggestions.

References

- Acevedo, P., Quirós-Fernández, F., Casal, J., & Vicente, J. 2014, Ecological Indicators, **36**, 594

- Baird, D. 1989, Cryptozoology, **8**, 43

- Bayanov, D. 2011, Learning from Folklore in Russian Hominology (Hancock House. Originally published on the Bigfoot Encounters website)

- Bayanov, D.. 2012, Historical Evidence for the Existence of Relict Hominoids in Russian Hominology (Hancock House. Originally published in The Relict Hominoid Inquiry, Idaho State University)

- Bayanov, D., & Bourtsev, I. 1976, Science and Religion, **39**

- Beer, A.-J., & Morris, P. 2004, *Encyclopaedia of North American Mammals* (Thunder Bay Press

- Beerkircher, L., Arocha, F., Barse, A., et al. 2009, Endangered Species Research, **9**, 81

- Bigfoot Field Researchers Organization. 2023, *Geographic Database of Bigfoot/Sasquatch Sightings & Reports*. https://bfro.net/GDB/

- Blight, E. A. 2005, Blight Investigations

- Bodley, J. H. 1988, Northwest science., **62**

- Brown, G. 1993, *Great bear almanac* (Lyons & Burford)

- M. M. Halpin & M. M. Ames

- Bryant, V. M., & Trevor-Deutsch, B. 1980, *in Manlike Monsters on Trial: Early Records and Modern Evidence*, ed. M. M. Halpin & M. M. Ames (Vancouver and London: University of British Columbia Press), 291–300

- Burton, J. A. 1998, Guide to Mammals of North America (Parkgate Books

- Cachel, S. 1985, Cryptozoology, **4**, 45

- Cartograhy Vectors. 2022, Combined US & *Canada with States & Provinces*. https://web.archive.org/web/20221128141742/ https://cartographyvectors.com/map/793-combined-us-canada-with-states-provinces

- Clark, J. D., Beckmann, J. P., Boyce, M. S., et al. 2021, American Black Bear (Ursus americanus)

- Colarusso, J. 1983, Cryptozoology, **2**, 90

- Coltman, D., & Davis, C. 2005, Trends in Ecology & Evolution, **21**, 60

- Costa, H., Foody, G. M., Jiménez, S., & Silva, L. 2015, ISPRS International Journal of Geo-Information, **4**, 2496

- Discovery. 2022, *Expedition Bigfoot Season 3 Episode 13 - A Massive Discovery*. https://web.archive.org/web/20221007103940/ https://www.travelchannel.com/shows/expedition-bigfoot

- Federal Bureau of Investigation. 2019, *FBI Records: The Vault - Bigfoot Part 01 of 01*. https://web.archive.org/web/20191220161650/ https://vault.fbi.gov/bigfoot/bigfoot-part-01-of-01/view

- Foxon, F. 2023a, bioRxiv, doi:10.1101/2023.01.07.523085

- Foxon, F.. 2023b, bioRxiv, doi:10.1101/2023.01.14.524058v1

- Foxon, F.. 2023c, bioRxiv, doi:10.1101/2023.01.14.524058v2

- Foxon, F.. 2023d, *Supplemental materials for paper: If it's there, could it be a bear?* https://doi.org/10.17605/OSF.IO/AV3G2

- Freeland, D., & Rowe, W. 1989, Skeptical Inquirer, **13**, 273

- Garshelis, D. L., Noyce, K. V., Ditmer, M. A., et al. 2021, Remarkable adaptations of the American Black Bear help explain why it is the most common bear: a long-term study from the center of its range

- Gorleri, F. C., & Areta, J. I. 2022, Ibis, **164**, 13

- Gorleri, F. C., Jordan, E. A., Roesler, I., Monteleone, D., & Areta, J. I. 2023, Ibis, **165**, 458

- Hart, H. V. 2016a, Relict Hominoid Inquiry, **5**, 8

- Hart, H. V.. 2016b, The Journal of Cryptozoology, **4**, 39

- Heaney, M. 1990, Cryptozoology, **9**, 52

- Herrero, S. 2018, Bear attacks: their causes and avoidance (Rowman & Littlefield

- Heuvelmans, B. 1986, Cryptozoology, **5**, 1

- Hristienko, H., & McDonald, J. E. 2007, Ursus, **18**, 72

- Kays, R. W., & Wilson, D. E. 2009, *Mammals of North America* (Princeton University Press), doi:10.1515/9781400833504

- Kelsey, P. 2022, *A Forensic Analysis of the Patterson-Gimlin Film*. https://web.archive.org/web/20221211054745/ https://adventuresinrediscovery.com/2022/06/13/a-forensic-analysis-of-the-patterson-gimlin-film/

- King, J. E., & Greenwell, J. R. 1983, Cryptozoology, **2**, 98

- Krantz, G. S. 1983, Cryptozoology, **2**, 53

- Lass, J., Bøggild, M. E., Hedegård, P., & Lefmann, K. 2021, Journal of Neutron Research, **23**, 69

- Lozier, J., Aniello, P., & Hickerson, M. 2009, Predicting the distribution of Sasquatch in western North America: anything goes with ecological niche modelling, doi:10.1111/j.1365-2699.2009.02152.x

- MacPhail, V. J., & Colla, S. R. 2020, Biological Conservation, **249**, 108739

- Mayor, A. 1991, Cryptozoology, **10**, 16

- McKinley, D. C., Miller-Rushing, A. J., Ballard, H. L., et al. 2017, Biological Conservation, **208**, 15, the role of citizen science in biological conservation

- Miller, D. A., Nichols, J. D., McClintock, B. T., et al. 2011, Ecology, **92**, 1422

- Molinari-Jobin, A., Kéry, M., Marboutin, E., et al. 2012, Animal Conservation, **15**, 266

- Munns, B., & Meldrum, J. 2013, The Relict Hominoid Inquiry, **2**, 41

- Munns, W. 2014, When Roger Met Patty (Self-published: CreateSpace)

- Napier, J. 1976, Bigfoot - The Yeti and Sasquatch in Myth and Reality (Sphere Books Ltd

- Nickell, J. 2013, Skeptical Inquirer, **37**

- Opit, G. 2017, Australian Zoologist, **38**, 430

- Paxton, C. G. M. 2009, Journal of Zoology, **279**, 381

- Pearce, J., & Ferrier, S. 2001, Biological Conservation, **98**, 33

- Pradhan, N. C., & Leung, P. 2006, Fisheries Research, **78**, 309

- Sanderson, I. T. 1967, Abominable Snowmen: Legend Come to Life (Chilton)

- Sawe, B. E. 2017, *Forest Land By Canadian Province And Territory*. https://web.archive.org/web/20221206040420/ https://www.worldatlas.com/articles/forest-land-by-canadian-province-and-territory.html

- Scheick, B., & Mccown, J. 2014, Ursus, **25**, 24

- Sillero, N., Arenas-Castro, S., Enriquez-Urzelai, U., et al. 2021, Ecological Modelling, **456**, 109671

- Somer, L. 1989, *Sasquatch Evidence: Scientific and Social Implications Symposium. Eight Annual Membership Meeting, International Society of Cryptozoology*, June 24–25, 1989, Washington State University, Pullman, Washington

- Spencer, R. D., Beausoleil, R. A., & Martorello, D. A. 2007, Ursus, **18**, 217

- Statistics Canada. 2022, *Population and private dwellings occupied by usual residents and intercensal growth, 1851 to 2021 censuses*. https://web.archive.org/web/20221128125153/ https://www150.statcan.gc.ca/t1/tbl1/en/tv.action?pid=9810001701

- Strain, K. M. 2012, Relict Hominoid Inquiry, **1**,

- Sumner, S., Bevan, P., Hart, A. G., & Isaac, N. J. 2019, Insect Conservation and Diversity, **12**, 382

- Sykes, B. C., Mullis, R. A., Hagenmuller, C., Melton, T. W., & Sartori, M. 2014, Proceedings of the Royal Society B: Biological Sciences, **281**, 20140161

- United States Census Bureau. 2021, *State Intercensal Tables: 2000-2010*. https://web.archive.org/web/20230327112644/ https://www.census.gov/data/tables/time-series/demo/popest/intercensal-

2000-2010-state.html

- USDA Forest Service. 2006, *Forest Inventory and Analysis Fiscal Year 2006 Business Report.* https://web.archive.org/web/20230314130101/ https://www.fia.fs.usda.gov/library/bus-org -documents/index.php

- Vonk, J., Jett, S. E., & Mosteller, K. W. 2012, Animal Behaviour, **84**, 953

- Winn, E. B. 1991, Cryptozoology, **10**, 55

Did the Smithsonian collect Thunderbird Bones in the Late 1800s?

Kevin J. Guhl

The Yup'ik of Alaska's Yukon Delta told of giant, man-eating eagles that nested in the mountains, and that a famed American explorer collected bones from one of these monsters. Was it true?

"From their perch on this rocky wall these great birds would soar away on their broad wings, looking like a cloud in the sky, sometimes to seize a reindeer from some passing herd to bring back to their young; again they would circle out, with a noise like thunder from their shaking wings, and descend upon a fisherman in his canoe on the surface of the river, carrying man and canoe to the top of the mountain. There the man would be eaten by the young thunderbirds and the canoe would lie bleaching among the bones and other refuse scattered along the border of the nest." - Edward W. Nelson, "The Last of the Thunderbirds" - Artwork ©2023 thunderbirdphoto.com

From 1881-1883, Norwegian ethnologist. Johan Adrian Jacobsen, journeyed across the Pacific Northwest of North America to collect indigenous artifacts on behalf of the Royal Berlin Ethnological Museum. As one author wrote, Jacobsen's mission was part of "a desperate effort" by museums and ethnographers to gather all they could of the handiwork of non-European peoples, before those societies were engulfed by modern culture. On his expedition to Alaska, Jacobsen kept finding that he had been beaten to the punch by Edward William Nelson, an American ethnologist and naturalist, who had procured many of the choicest artifacts and specimens for the Smithsonian Institution during his own assignment in Alaska, with the U.S. Army Signal Corps. from 1877-1881. Nelson was stationed at Fort Saint Michael in Norton Bay, located on Alaska's central west coast. As incredible as it sounds, among the specimens Nelson collected were supposedly the bones of a giant, man-eating bird known to the native Alaskans.

"From their perch on this rocky wall these great birds would soar away on their broad wings, looking like a cloud in the sky, sometimes to seize a reindeer from some passing herd to bring back to their young; again they would circle out, with a noise like thunder from their shaking wings, and descend upon a fisherman in his canoe on the surface of the river, carrying man and canoe to the top of the mountain. There the man would be eaten by the young thunderbirds and the canoe would lie bleaching among the bones and other refuse scattered along the border of the nest." - Edward W. Nelson, "The Last of the Thunderbirds" - Artwork ©2023 thunderbirdphoto.com

In September 1882, Jacobsen was trekking through the Yukon-Kuskokwim Delta in western Alaska, in the vicinity of Mission (today called Russian Mission). On the hunt for artifacts, Jacobsen visited a place called Nunalinak and purchased some stone pieces. A couple of hours later, he arrived at a spot of great significance to his native companions. Jacobsen wrote (translated from German):

Johan Adrian Jacobsen. Public domain, via Wikimedia Commons.

"Toward seven o'clock, we came to a mythological site where my Indians insisted on taking a side trip to see the place where a giant bird lived, who attacked adults and carried them to his nest as food for his children. I was assured that on a nearby mountaintop were the remains of the nest and even bones that I would want to collect. I was convinced and pitched our tent for the night, then engaged a young Indian to take me to the next village, the site of the nest. By a forced march we reached the 1,500— 1,800-foot-high mountaintop, which we could recognize from the distance as a steep tower-like structure. Because of my past climbing experience I succeeded in ascending the thirty-foot high rock that was supposed to be the nest of the giant bird, but there was not a sign of anything in it. However, the view of the landscape from here was worth the effort. An endless fine veil of golden fog was lying on the land between the

Yukon and the Kuskowkim Rivers, while the small waterways and lakes lay on the land like a net".

As darkness began to settle in, Jacobsen and his party descended the mountain and headed back to their boat on the bank of, presumably, the Yukon River. He continued:

"My presence had become known and a number of people gathered and discussed my trip up the mountain with great eagerness. Their faith in the presence of this bird was not shaken by my not finding it, and some of the older ones insisted that in their youth they had seen the giant bird with their own eyes. It was also said that an Eskimo found the leg bones of an usually large bird and gave them to a traveling scholar". This is supposed to have been Mr. Nelson, who was collecting for the Smithsonian Institution.

This is a stunning claim that begs the question of whether Nelson indeed procured the remains of an extraordinarily large bird for the Smithsonian. While the answer might be more complex, historical records bear out a skeleton of truth beneath the story.

Edward W. Nelson in Alaska. Public Doman, via Wikimedia Commons.

Nelson published the findings of his expedition in "Report Upon Natural History Collections Made in Alaska Between the Years 1877–1881" (1887) and "The Eskimo

About Bering Strait" (1900). In both volumes, Nelson shared stories of a monster bird very similar to Jacobsen's account, alternatively called Mutughowik (Mû-tûgh'-o-wĭk) and "Tĭñ-mĭ-ûk'-pûk, the great eagle (Thunderbird)." Later authors would render Tinmiukpuk as "Tengmiarpak."

Nelson described *"the belief that long ago the eagles were larger and fiercer than they now are."* In both of the reports on his Alaska expedition, Nelson included slightly different retellings of a tale called "The Last of the Thunderbirds." He wrote:

"The story is current among the Eskimo along the Lower Yukon and neighbouring coast: In ancient times there were eagles of tremendous size frequenting the tops of the highest mountains in the interior and preying upon whales and full-grown reindeer, and even upon men. A volcanic crater of very regular outline, situated upon the summit of a mountain near the Lower Yukon, was pointed out to me as the nest of the ancient Mutughowik. Around the rim of the crater are differently-colored stones, which, the natives claim, were gathered by these birds to ornament their nest. When the birds sat here, overlooking the Yukon on the one side and the sea far away to the horizon on the other, their screams could be heard for miles, and many luckless creatures were caught in their talons and carried swiftly to their eyrie, and there torn into fragments to be devoured. Year after year these birds remained, until men were afraid to go out on the broad bosom of the Yukon for fear of being caught by these evil guardians of the mountains overlooking their village. Each year the young were raised and flew away, none knew whither; so that never more than two old birds inhabited the mountains."

Nelson relayed a native tale about a renowned hunter, whose wife was taken by one of the giant eagles, whose wings shook like thunder. It snatched her while she was filling a bucket of water in the river and carried the woman off to feed to its nestlings. The mourning hunter grabbed his best bow and arrows, intent on exacting revenge. He scaled the mountain and, at the summit, discovered an enormous nest. *"All about were strewn fragments of men and animals, among which were seen the frames of many kayaks,"* wrote Nelson. The raging female swopped down toward the hunter and he fired an arrow into her side, sending the bird tumbling down the mountain. He then turned his deadly bow on her brood. His vengeance seemingly fulfilled, the hunter was suddenly assaulted by the even more fearsome male eagle. While many arrows found their mark, the wounded eagle escaped and flew off to the north, the direction young eagles would fly every fall. With the nesting pair gone, the monster birds were never seen again and people were able to resume hunting along the Yukon River without fear of being carried off in their talons. Nelson concluded:

"The villagers afterwards visited the nest with their deliverer and found many relics of friends who had perished, and it was only a few years ago that the remains of the kyaks [sic] were still to be seen about the nest." This story is implicitly believed by the natives of the Lower Yukon and adjacent sea-coast, and the Bald Eagle is known by the name which they apply to the bird of their legend.

Stories of heroes who ascend to the nests of monster birds and slay them exist in indigenous cultures throughout North America. Variations exist among peoples such as the Cherokee, Navajo, Northern Shoshone, Wampanoag, Washo and Yavapai. While a similar motif has been attributed to the Illini in the infamous tale of the man-eating Piasa Bird of Illinois, it is likely this tale was actually the invention of author John Russell in 1836. The Tengmiarpak shares similarities with the version of the Thunderbird described by the Hoh and Quileute in the Pacific Northwest, that of a giant raptor that would carry off whales to its nest to feed its young.

Both Jacobsen and Nelson offer intriguing hints about the identity of the Alaskan mountain on which the monster birds nested, without providing exact names or coordinates. However, there are a few candidates...

Jacobsen was traveling along the Yukon River south of Mission, also the site of a native village called Ikogmiut. Starting in the mid-19th century, the settlement had hosted a Russian-American Company fur trading post, a Russian Orthodox Church mission and, after the United States' acquisition of Alaska in 1867, and Alaska Commercial Company store. Nunalinak is noted as being near the giant bird nest, but mention of this settlement, let alone its location, is elusive outside of Jacobsen's book. The morning after his journey up the mountain, Jacobsen wrote that his next stop was the village of Ka-krome, where he described *"the remainders of houses for about four English miles"* along the riverbanks. This is another village lost to time, although Kenneth L. Pratt, Alaska-based anthropologist and ethnohistorian with the U.S. Bureau of Indian Affairs, suggested Ka-krome was *"either at or in the immediate vicinity of Unglurmiut"* (unglu meaning "nest"), located on the Lower Yukon River about 20 km or 12.4 miles southwest of Russian Mission. Possibly abandoned due to civil war, traces of the village were no longer visible by the late 1920s, likely overgrown with dense brush and grass. It might also be noted that smallpox took a devastating toll on native Alaskans; an epidemic in the 1830s wiped out entire villages and killed two-thirds of the native population in the lower Yukon.

The tallest formation in the area that Pratt pinpoints for Unglurmiut, and possibly Ka-krome, is Bend Mountain. It possesses an elevation of 1,535 feet, within Jacobsen's estimate of 1,500-1,800 feet. Positioned on the north bank of the Yukon River, the flat-topped mountain is located 11 miles southwest of Russian Mission, in the Kusilvak Census Area. Bend Mountain obtained its name from riverboat pilots because it indicated the downstream approach to a sharp bend in the Yukon River called Devils Elbow. This would appear to be a reasonable candidate for the mountain that Jacobsen scaled in search of the giant eagle's nest.

However, there is another nearby mountain associated with such legends. Pilcher Mountain, named for miner G.M. Pilcher, is another flat-topped formation and reaches 1,948 feet. It is positioned four miles northeast and overlooking the city of Marshall. Mount Pilcher is located approximately 23 miles northwest of Bend Mountain, along the Yukon River and about 24 miles west and slightly north of Russian Mission. In the

Bend Mountain, shown on Google Maps. Used here on a Fair Use, non-commercial and educational basis.

Mount Pilcher, shown on Google Maps. Used here on a Fair Use, non-commercial and educational basis.

language and traditions of the Yup'ik people, who inhabit the Yukon–Kuskokwim Delta, the top of Mount Pilcher is said to have once housed the nest of the Tengmiarpak, the "mythical 'thunderbird.'" In the Yup'ik language, Tengmiarpak can also refer to bald and golden eagles. According to local legend, Pilcher Mountain is also reputedly home to an elfin people called the Inukin.

While Nelson and Jacobsen recorded what is apparently the same cultural tradition of giant eagles, Nelson's description of the nest location differed somewhat.

Nelson wrote that the native Alaskans pointed out *"the crater of an old volcano as the nest of the giant eagles, and say that the ribs of old canoes and curiously colored stones carried there by the birds may still be seen about the rim of the nest."* The birds nested on a *"mountain top overlooking the Yukon river near Sabotnisky. The top of this mountain was round, and the eagles had hollowed out a great basin on the summit which they used for their nest, around the edges of which was a rocky rim from which they could look down upon the large village near the water's edge,"* wrote Nelson.

Sabotnisky is an alternative name for Uglovaia (aka Ooglovia, Ouglovaia or Uglivia), a village once located on the right bank of the Yukon River at or near Marshall. This could suggest that the native Alaskans were referring to Mount Pilcher, although it does not appear to have been volcanic. There are volcanic fields further west in the Yukon region—the Ingakslugwat Hills north of Baird Inlet, the Kusilvak Mountains adjacent to Nunavakanuk Lake, and Nelson Island (yes, named after that Nelson).

In 1985, Pratt attempted to locate the nest of the Tengmiarpaks while conducting archaeological work in the lower Yukon. His goal in finding the site was to verify it as an Alaska Native historical place under Section 14(h)(1) of the Alaska Native Claims Settlement Act (ANCSA) of 1971. An application had been filed, citing *"an unnamed volcanic crater situated on a tundra plain north of the Yukon River mouth, near the western base of the Nulato Hills."* Near three other volcanic craters, this area is north of Russian Mission and closer to St. Michael and Stebbins on the coast of Norton Sound. Assisted by interpreters who spoke Central Yup'ik, Pratt interviewed a variety of native sources. What he soon learned was that there are disagreements and contradictions about the location of Tengmiarpak nests, as well as whether those nests had housed the giant birds or their smaller eagle cousins. Pratt ultimately amended the ANCSA application to specify a mountain that was *"on a rocky cape of land overlooking the Bering Sea, about three miles west of Stebbins and twenty miles northwest of the site location reported in the application."* Reaching the site by helicopter, a native elder guided Pratt to envision how a giant nest would have appeared at that location in the distant past, littered with the bones of beluga whales that had been fed to the young, and how the shape of the land represented the Tengmiarpak. While the search had been partially inspired by Nelson's account, Pratt believed that the mountain in Nelson's story was more likely situated on the lower Yukon River, between Russian Mission and the village of Holy Cross.

So, we return to our most pressing question—Did Nelson collect "Thunderbird" bones for the Smithsonian? While Nelson delved into the native tales of giant eagles in both of his works about Alaska, he made no mention of acquiring the remains of one of these monster birds. It's hard to imagine he would omit such a supporting fact, especially since he elaborated on other artifacts that represented Alaskan Thunderbird traditions.

Today, the Smithsonian provides online access for the public to search for and view the wealth of artifacts and specimens that Nelson collected during his fruitful tenure in Alaska. Nelson gathered samples of several bird species, although abnormally large bones are not catalogued among them. The artifacts Nelson acquired include small items made from bird bones such as needle cases and snuff tubes. Some of the native Alaskan pieces Nelson sent to the Smithsonian depict Thunderbird traditions. One excellent example is a Yup'ik tool box with painted figures on its lid that include Thunderbirds grasping victims such as a caribou, a whale and a man in a kayak in their talons.

This Yup'ik seal-shaped tool box was collected by Edward W. Nelson in the village of Pastolik near the mouth of the Yukon River in 1879. According to the Smithsonian, Yup'ik men owned tool boxes which were often illustrated with hunting art and imagery such as Thunderbirds, which are depicted here carrying off victims such as a whale, a caribou and a man in a kayak. Photograph courtesy of the Smithsonian.

In recent years, the Smithsonian Institution Archives enlisted digital volunteers to review Nelson's handwritten field notes from Alaska and other regions of North America and transcribe them so the content could be published in a more accessible form online. This is where we finally locate a clue that could explain why Jacobsen was informed that Nelson has acquired the remains of a giant bird.

Nelson recorded the relevant passage during his Fort Saint Michael assignment on Oct. 15, 1880:

"I bought a fragment of buffalo horn from a native. It had washed ashore from the sea at Unalakleet. The natives say it is part of the claw of a large kind of eagle which formerly lived there."

St. Michael, 1901. National Archives and Records Administration, Public domain, via Wikimedia Commons.

Nelson and Jacobsen conducted their research in the same areas of the Yukon–Kuskokwim Delta along the coast of the Bering Sea, Jacobsen following Nelson's tenure by only a few months. They likely encountered many of the same native people, and they were told similar stories about the giant, man-eating eagles that once descended from the mountaintops to reign down terror on the populace. It's not inconceivable that Nelson's acquisition of a "claw" from one of these monster birds is what evolved, either through retelling or Jacobsen's interpretation, into Nelson obtaining "leg bones" of one of these giant eagles.

So, does the Smithsonian have Nelson's "buffalo horn" fragment in its collection? When I inquired, Smithsonian archivist, Tad Bennicoff, smartly pointed out that a muskox horn catalogued among the Institution's accessions from Nelson might be the very same specimen. Both large bovines, the bison/buffalo and muskox are often compared. And there is indeed a visual similarity between an eagle talon and the tip of a muskox horn. Interestingly, the Alaskan population of muskox were wiped out in the late 19th or early 20th century due to excessive hunting and possibly climate change. Nelson wrote in 1900 that the muskox had been extinct "for ages" in the lower Yukon. However, muskox have since been reintroduced to Alaska.

Three Wet Muskox, Bering Land Bridge National Preserve, CC BY 2.0, via Wikimedia Commons.

Darrin Lunde, collections manager for the Division of Mammals at the Smithsonian, helpfully agreed to retrieve the muskox horn from offsite storage and provide photographs and measurements of the fragment. Unfortunately, he found that although the horn had been catalogued by the Smithsonian in 1885, it was not present in any of the drawers holding Alaska muskox specimens, wasn't mentioned on the drawer labels, and had never been reported missing. So, its whereabouts remain a mystery, and it might very well have been lost sometime in the past 138 years. In any case, I greatly appreciate the Smithsonian Institution staff's perpetually pleasant and accommodating nature.

(The absolute last thing I want this article to do is spin yet another conspiracy theory that the Smithsonian is hiding discoveries that don't mesh with mainstream science. It's a trend prevalent enough that the Smithsonian has felt compelled to dispel rumours like its purported discovery of Egyptian-like ruins in the Grand Canyon and its alleged destruction of thousands of skeletons of giant humans discovered across America.)

Several questions and loose ends remain: What is the genesis of native stories about giant eagles in Alaska? Did Nelson or some other, less prominent, explorer collect the massive leg bones of such a creature? Could they have been the remnants of a large bird that once occupied the region but has since disappeared? Perhaps they were leg bones from the California condor, which is known to have occurred as far north as British Columbia during the Pleistocene to Late Pleistocene...

To the people to whom the Tengmiarpak belongs, though, these questions might just not matter. "*In 1985, Yup'ik residents of this region still firmly believed in the physical existence of tengmiarpak nest sites,*" wrote Pratt. But he concluded that locating physical evidence of a nest was not viewed as necessary among the Yup'ik. "*To those individuals, firsthand accounts and traditional Native history are sufficient to prove the existence of these creatures.*"

In a sense, Nelson did collect the remains of a Thunderbird in Alaska during the late 1800s on behalf of the Smithsonian Institution, much like Jacobsen conveyed. Perhaps the details were a bit fluid, but Nelson did send home a Tengmiarpak "claw," even if his scientific mind quickly identified it as part of a muskox horn. But more importantly, Nelson and Jacobsen collected something arguably more valuable—a history of living, breathing traditions, both fearsome and beautiful, as they existed in a moment in time when Alaska was less infiltrated by the modern world.

—Kevin J. Guhl

SOURCES:

- "American Commercial Company." Wikipedia, https://en.wikipedia.org/wiki/Alaska_Commercial_Company. Accessed 6 Jan 2023.

- D'Elia, Jesse. California Condors in the Pacific Northwest: Integrating History, Molecular Ecology, and Spatial Modelling for Reintroduction Planning. 2014. Oregon State University, PhD dissertation.

- "Edward William Nelson." Wikipedia, https://en.wikipedia.org/wiki/Edward_William_Nelson. Accessed 26 Dec. 2022.

- "Edward W. Nelson - Alaska, October 6 - December 6, 1880." Smithsonian Institution Archives, https://transcription.si.edu/project/27107. Accessed 15 Jan. 2023.

- "Ethnology: Tool Box." Smithsonian National Museum of Natural History, http://n2t.net/ark:/65665/3c46ed6a9-bfe3-4459-bf8c-31ca703da690. Accessed 15 Jan. 2023.

- "'Great Death' Smallpox Epidemic in Alaska." Investing in Native Communities, https://nativephilanthropy.candid.org/events/great-death-smallpox-epidemic-in-alaska/. Accessed 6 Jan. 2023.

- "Ingakslugwat Hills." Wikipedia, https://en.wikipedia.org/wiki/Ingakslugwat_Hills. Accessed 7 Jan. 2023.

- Jacobsen, Johan Adrian. Alaskan Voyage 1881-1883: An Expedition to the Northwest Coast of America, from the German Text of Adrian Woldt. Translated by Erna Gunther, The University of Chicago Press, 1977.

- "Johan Adrian Jacobsen." Wikipedia, https://en.wikipedia.org/wiki/Johan_Adrian_Jacobsen. Accessed 26 Dec. 2022.

- "Kusilvak Mountains." Wikipedia, https://en.wikipedia.org/wiki/Kusilvak_Mountains. Accessed 7 Jan. 2023.

- "Muskox." Wikipedia, https://en.wikipedia.org/wiki/Muskox. Accessed 15 Jan. 2023.

- "Native American Legends: Piasa Bird (Piesa)." Native Languages of the Americas, http://www.native-languages.org/piasa.htm. Accessed 10 Jan. 2023.

- Nelson, Edward W. Report Upon Natural History Collection Made in Alaska Between the Years 1877 and 1881, edited by Henry W. Henshaw, Washington, Government Printing Office, 1887.

- Nelson, Edward William. The Eskimo About Bering Strait: Extract from the Eighteenth Annual Report of the Bureau of American Ethnology. Washington, Government Printing Office, 1900.

- Orth, Donald J. Dictionary of Alaska Place Names. United States Government Printing Office, 1967.

- Pierce, Richard A. Review of Alaskan Voyage 1881-1883. An Expedition to the Northwest Coast of America, by Johan Adrian Jacobsen, translated by Erna Gunther from the German text of Adrian Woldt. The Journal of San Diego History: San Diego Historical Society Quarterly, Summer 1978, https://sandiegohistory.org/journal/1978/july/br-alaskan/. Accessed 26 Dec. 2022.

- Pratt, Kenneth L. "Deconstructing the Aglurmiut Migration: An Analysis of Accounts from the Russian-America Period to the Present." Alaska Journal of Anthropology, vol. 11, nos. 1&2, 2013, pp. 17-36.

- Pratt, Kenneth L. "Legendary Birds in the Physical Landscape of the Yup'ik Eskimos." Anthropology and Humanism, vol. 18, no. 1, 1993, pp. 13-20.

- Reagan, Albert B. "Some Myths of the Hoh and Quillayute Indians." Transactions of the Kansas Academy of Science," vol. 38, March 28-30, 1935, pp. 43-85.

- Reuters Fact Check. "Fact Check-Claims that the Smithsonian Destroyed 'Thousands of Giant Skeletons' are Many Years Old and Satirical." Reuters, 4 Aug. 2022, https://www.reuters.com/article/factcheck-skeletons-smithsonian/fact-check-claims-that-the-smithsonian-destroyed-thousands-of-giant-skeletons-are-many-years-old-and-satirical-idUSL1N2ZG1I0. Accessed 15 Jan. 2023.

- Rhodes, Jesse. "175 Years of the Smithsonian's Most Untrue Stories." Smithsonian Magazine, 31 Aug. 2009, https://www.smithsonianmag.com/smithsonian-institution/urban-legends-about-the-smithsonian-135407460/. Accessed 15 Jan. 2023.

- "Russian Mission, Alaska." Wikipedia, https://en.wikipedia.org/wiki/Russian_Mission,_Alaska. Accessed 6 Jan. 2023.

- Swancer, Brent. "Odd Encounters with the Mysterious Little People of Alaska." Mysterious Universe, 15 Dec. 2017, https://mysteriousuniverse.org/2017/12/odd-encounters-with-the-mysterious-little-people-of-alaska/. Accessed 7 Jan. 2023.

- "Tengmiarpak." Yup'ik Eskimo Dictionary. 2nd ed. Compiled by Steven A. Jacobson, vol. 1, 2012.

- "Yupik Peoples." Wikipedia, https://en.wikipedia.org/wiki/Yupik_peoples. Accessed 6 Jan. 2023.

THE MAN EATERS OF BORNEO
Richard Freeman

The island of Borneo is a hotspot for man-eating crocodiles. Margaret Brook was the wife of Sir Charles Brooke, the second white Raj of Sarawak. In her 1913 book *"My Life in Sarawak"* she writes of a huge crocodile that haunted a local river.

> *"A great many years ago, before Kuching became as civilized as it is now, and when it had few steamers on the river, an enormous crocodile, some twenty feet in length, was the terror of the neighbourhood for three or four months during the north-east monsoon - the rainy season of the country. Our Malay quartermaster on board the Heartsease was seized by this monster as he was leaving the Rajah's yacht to go to his house, a few yards from the bank, in his little canoe. It was at night that the crocodile seized him, the canoe being found empty the next morning. Although no one had actually witnessed the calamity, it was certain the poor man had been taken by the monster. This was his first victim, but others followed in quick succession.*

Island of Borneo

BRUNEI   MALAYSIA   INDONESIA

The crocodile could be seen patrolling the river daily, but it is very difficult to catch or shoot such a creature. At length the Rajah, becoming anxious at the turn affairs were taking, issued a proclamation offering a handsome reward to any one who should succeed in catching the crocodile. This proclamation was made with as much importance as possible. The executioner, Subu, bearing the Sarawak flag, was given a large boat, manned by twenty paddles, painted in the Sarawak colours, and sent up and down the river reading the proclamation at the landing-stages of Malay houses.

Looking from my window one morning, I saw the boat gaily decorated and looking very important on the river, with the yellow umbrella of office folded inside and the proclamation from the Rajah being read. A few yards behind the boat I imagined I could see, through my opera glasses, the water disturbed by some huge body following it. The natives *had noticed this too, and it was absolutely proved that wherever the boat went up or down the river, the monster followed it, as if in derision of the proclamation."*

The best known Bornean man-eater was christened Bujang Senang, the Happy Bachelor. His story runs like the plot of a horror novel or action movie. It began on the 26th of June, 1982 in the Batang Lumpar river, Borneo.

Bangan Anak Pali, an Iban farmer and his brother Kebir, had decided to collect shrimps in the river. Bangan had recently received a letter confirming his appointment as a chieftain of the Ibans living around the Tanjung Bijat area.

Whilst wading in the river, Batang felt that he was suddenly standing on a large log. It was then that the water exploded. The man was swept off his feet by the crocodile's huge tail and seized by it's huge jaws. Kebir said the animal was longer than their boat and had a distinctive white back.

A day later the crocodile was seen flipping Batang's mutilated body about like a rag doll. Bujang Senang devoured most of the man's body. Only his head and upper torso were retrieved.

In another river, the Tebu, an 86 year old Iban man, Abang Anak Gelayah, was taken by a massive crocodile that rammed and overturned his boat. His daughter-in-law and

granddaughter managed to swim to the bank but the old man was grabbed and dragged under.

Soon after the local police launched Operation Buaya Ganas (Operation Ferocious Crocodile). It was led by Sarawak Commissioner of Police, Datuk Seri Yuen Leng, and involved men armed with modified bear traps, harpoons, guns and grenades. A group of medicine men from Indonesia joined the hunt. One of these claimed to have harpooned Bujang Senang from a boat, but he was so powerful that he dragged the vessel against the current, almost sinking it and broke free, dislodging the harpoon.

The crocodile took bait such as monkeys from hooks but never got caught. At one point ducks with grenades attached were employed! Bujang Senang declined such delicacies. Hunters threw spears at the monster but they simply bounced off his scales. On another occasion, two grenades were hurled into the water but the killer croc just sank and vanished.

On November 3rd, a crocodile took a monkey bait attached to floats. In a one hour struggle to land the beast, it almost capsized the hunter's boat. They shot at it multiple times and threw no less than five grenades into the water. When finally subdued, the creature was found to be fifteen feet long, too small to be Bujang Senang.

It would be two years before Bujang Senang struck again. On September 27th, 1984, a 51 year old Iban man Badong Anak Apong was fishing for shrimps close to the sight of the first attack. The crocodile rammed his boat and capsized it. It grabbed Badong and dragged him to a muddy bank whilst thrashing him violently. Five men and the river bank witnessed it. Running to the tribal longhouse they alerted six hunters who, armed with rifles returned to the river to see the huge crocodile shaking and tossing the man's body. Despite shooting it twice they were unable to get the crocodile to let go. The victim was utterly devoured. Again the distinctive white back was reported. A large reward was offered for anybody able to kill or catch Bujang Senang, but nobody claimed the cash. All went quiet for the next few years.

Five years later, on February 20th, 1989, farmer Berain Anak Tunggling was repairing his boat in shallow water when the white backed crocodile dragged him off his boat and into the water.

A 70 year old man named Tuah Anak Tunchun came forward and said that the same white backed crocodile had eaten his brother Inchi, in 1962, at Sungei Selumbang. Others claimed to have seen it in the company of an even larger black (in reality very dark green) crocodile. It was postulated that five other people devoured between 1965 and 1979 were victims of Bujang Senang.

On May 21st, 1992, 30 year old Dayang Anak Bayang was wading across a tributary to get to hunt. Unknown to her Bujang Senang was lying in wait on the bed of the tributary. Her mother, Umai saw Dayan suddenly vanish under the water. Then a huge

Legenda Bujang Senang

JUN. 1992

Oleh Normore Mullie

Pali iaitu abang kepada Bangan, adiknya sedang menjala udang di sungai sedalam tiga kaki apabila dia tiba-tiba mendengar bunyi air bercepuk yang sungguh kuat.

"Saya duduk di atas perahu yang diikat dengan tali dan ditarik oleh Bangan. Tiba-tiba adik saya berkata dia berasa terpijak di atas sebatang kayu besar terbenam dalam lumpur berhampiran tebing sungai.

"Selepas itu air berkocak dan Bangan dipukul seekor buaya besar sebelum terhumban ke sungai. Saya cuba mencari Bangan dan ketika itu ternampak bayangan seekor buaya mendekati perahu.

"Ketika itu, saya tidak rasa takut sebaliknya terjun ke sungai dan memeluk ekor buaya yang begitu besar untuk

AKHIRNYA... buaya putih dipercayai Bujang Senang yang mengorbankan ramai mangsa. Adik Dayang, Enie, salah seorang mangsanya berdiri dua dari kiri.

DITOREH... 'Buj mengikut kepe

Selepas penduduk berukuran kira-kira tujuh ... itu, ialah buaya ka...

sang tetapi is bernama Dayan

Rangka Kepala Bujang Senang

BORNEO ORACLE

crocodile broke the surface holding the woman in it's jaws. Umain flailed at the monster with a tree branch to no avail. With a flick of his tail the beast carried Dayan away into the water.

Several people who were nearby heard the commotion and ran to help. Saperie Anak Kiang shot at the beast, but missed. Dayang's brother, Enie Anak Bayan, was fishing downriver and was informed of the attack. Armed with a shotgun he paddled upriver and saw the crocodile with his sister in it's mouth. He shot it twice to no effect.

News reached the longhouse were Dayang lived, and by noon 26 people armed with guns and spears had gathered at a deep pool where Bujang Senang was thought to be lurking. Enie found his sister's body in the reeds and it was retrieved and put on a boat. Umain fainted at the sight.

Suddenly the crocodile emerged to reclaim it's prey. Enie was only ten feet away and fired his shotgun, hitting it in the eye. It thrashed violently and sunk. Men threw spears but missed. Mandauu Anak Tabor, a neighbour who had joined the hunt managed to throw a spear into the monster's back. The crocodile made its way upriver with the spear still jutting from it's back. It smashed it's way through a fence erected across the river. And it was shot again and again. Bujang Senang continued upriver until he found his way blocked by a fallen tree, and turned back and swam to the pool. Mandauu bravely tried to force the spear deep into the giant beast's back, but it bent on the iron hard scales.

Jaws agape, the crocodile lunged torward Mandauu, Enie and another man, Sidi Anak Iman, shot into Bujang Senang's mouth from point blank range. The killer reared up out of the water, standing on it's tail. Crashing back down it bit madly at logs and debris in the water, before sinking it's teeth into a tree stump and expired.

It took 24 men to haul the massive body onto a boat. The Happy Bachelor was measured at 19 feet 3 inches and tipped the scales at over a ton. He was a big crocodile but by no means the biggest.

The 'black' crocodile sometimes seen with Bujang Senang was even bigger according to witnesses. Pandi Anak Lai said...

> *"I have seen a black crocodile which is at least 35 feet long, along the Batang Lumpar."*

Bujang Senang may have bit the dust, but there were other killer crocodiles stalking the rivers of Borneo.

At about 7.30am, on January 1st, 1993, 12-year-old Masri Bujang, was sitting near the back of his father's sampan on the upper reaches of the Sungei Samarahan. His father was casting a net and his younger brother holding up a kerosene lamp.

Suddenly, an 18 foot crocodile emerged from the water and grabbed the boy. Masri's father, Bujang Amet, bravely leapt into the water to save his son but the crocodile was so powerful it simply dragged the boy away with no effort.

Sungei Samarahan is a river well known for large, man-eating crocodiles. A crocodile hunter called Talip was employed to deal with the man-eaters. During a five month period in 1917, he caught 42 of them along the river, the biggest being 21 feet long.

On October 28th, 1993, 42-year-old Abang Saperi bin Saad, was bathing in the river at his village of Kampung, Melaya, when a huge crocodile grabbed him. His screams brought neighbours with torches, but by the time they got there the reptile had dragged

his prey away. The people dubbed this new man-eater Bujang Samarahan.

A hunt was organized and the men involved said they came across a crocodile as 'thick as a coconut tree trunk,' but were afraid to use their guns as they thought the weapons would have little effect.

A 14 foot crocodile was later killed by the police, but it was clearly too small to be Bujang Samarahan.

After a long day's work, farmer Menggong Anak Madon, 49, his wife Gidong Anak Laking, and their 13-year-old son Stephen, were taking a bath in the Batang Sererap, a tributary of the Batanf Lupar, some 20 minutes walk from the village longhouse at Tanjung Marwar. It was about 5p.m on March 2nd, 1995.

Gidong heard a splash and saw her husband vanish beneath the surface.

> *"It happened so fast. I heard my husband shout for help, saying I am dying, and the water was churning as if a struggle was going on underwater. Then it was over. I cannot believe he is dead."*

A search party saw an eagle circling about the water about half an hour's journey away from the farm the next day. They found what was left of Menggong. The crocodile had eaten his whole lower body and one arm. Only his head, upper torso and one arm were left.

On November 22, 1993, two police marksmen, Awi anak Brodie and Busan anak Launaka, together with six other hunters, came across a huge crocodile. They hurled four grenades at it and fired 14 rounds into it with high powered rifles. The struggle took half an hour but finally the great beast lay dead. It was hauled onto a sampan and later had it's stomach opened. Inside it was a watch, that Abang Saperi's widow identified as belonging to him.

But other giant man eaters were at work in the area.

At around 6.30am on Christmas Day, 1996, Kayak Anak Entili a 27-year-old Iba housewife, was bathing in a tributary of the Sungei Tisak. Suddenly, in front of several witnesses a huge crocodile emerged from the water and grabbed her. A police boat was dispatched to the area and a search party that included Kayak's husband began to look for her.

> *"On Boxing Day at 2.15p.m her mangled remains were discovered. The crocodile had eaten her head, arms and legs."*

In March of 2010, a man named Sahar from Manubar Village, Sandaran District, Kalimantan, was fixing his boat in the Manubar river, when he was grabbed and eaten by a huge crocodile. The animal was later shot by members of the Indonesian navy.

Police Chief, Andi Razak, said that the creature was so big that it almost sank the fishing boat the body was being transported on and had to be moved to a bigger boat. This in turn was pulled by two other boats and took the crocodile to Sangkulirang pier. It then took 120 men to haul it to the front of the police headquarters.

Cutting open the belly, the police found a buffalo's leg, and Sahar's body with it's spine snapped and ribs crushed. The crocodile's body was preserved as an exhibit, by the Head of the Sangkulirang Health Center, Dr. Markus Sambo. Andi Razak said that many people had contacted him asking if they could have the animal's 'tangkurs' (genitals).

The crocodile was measured at 20 feet.

"A woman named only as Fatimah was killed and eaten by a 19 foot long crocodile in North Kalimantan in July of 2022. The 45 year old had been fishing at night in a river on Tibi Island in Bulungan Regency. Apparently she had bee throwing food into the water to attract fish. He friends heard her scream as the reptile bit into her."

Police and residents began a search operation, and after just a few hours they found her head and body parts. Next day the crocodile was shot, and the rest of her remains were found in it's stomach.

It is not only natives that fall prey to the jaws of Indo-Pacific crocodiles in Borneo. In 2002, Richard Shadwell, a 35-year-old musician from Sutton, in Surrey, was eaten by a crocodile in the Simpangkanan River in Tanjung Puting National Park. In an act of extreme foolishness, that would be worthy of a Darwin Award, Richard was swimming behind a boat in the crocodile infested river.

"Shortly after Mr Shadwell plunged into the river, I saw a black crocodile devour him, then he vanished," a guide called Jeki, told the Indonesian News Agency.

In 2022, the *Daily Star* reported the capture of a 26 foot man-eating crocodile in the Semaja River in the Nunucan Regency, part of North Kalimantan on Borneo. The crocodile had killed and eaten Samsul Bahri, 45, as he fished for shrimp on the 19th of July.

Police and villagers captured and sedated two crocodiles 14 and 16 feet long. The crocodiles were induced to vomit, but no human remains were found in their stomachs. Later the 26 foot giant was snared. Again this crocodile was sedated and induced to vomit. The contents of the creature's stomach contained partially digested human remains that were identified as Samsul. No further details of this case have come to light at the time of writing.

Frazier Valley: Canada's other land of Mystery.

David Scott

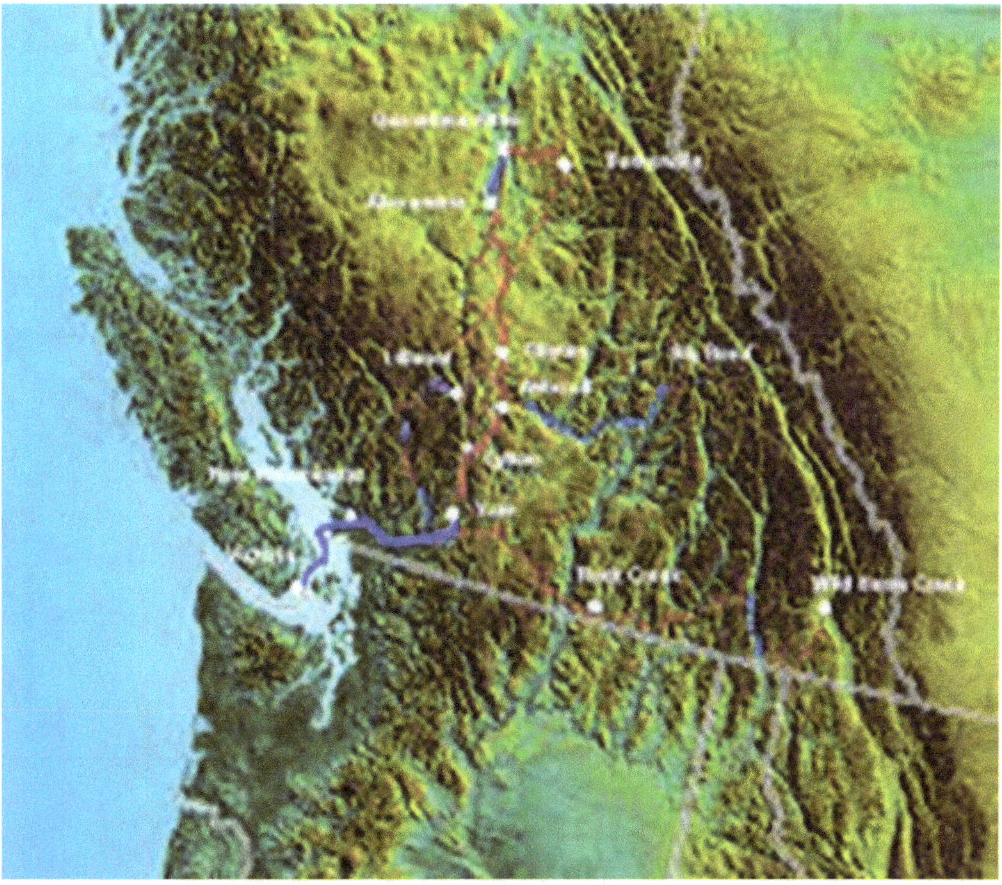

As head of CFZ Canada, I wanted to mount an expedition into the Nahani Valley of the North West Territories. After doing a proposed cost evaluation, I found the trip would have been 1) Cost prohibitive, and 2) because of its status and only being able to obtain set tour operations that would hinder our study, I nixed the whole idea and began looking for alternative for 2024.

With a study in Northern Ontario slated and in the books for 2024, my mind once more drifted back to my first idea, and found the Frazier Valley, in British Columbia, Canada. With roadways and access to major cities within a few hours drive, it met all the criteria of a proper study area, with a history rich in folklore, cryptid sightings and rare species, and perhaps new species of wildlife, plants and the things we all cherish so much.

Let me share some of the history and reports of the Valley. Clear your mind, and picture yourself with us, in the proposed expedition for 2025. Please comment any ideas you'd like us to look into while there.

The Frazier Valley encompasses the upper and lower valleys, beginning 100kms inland, at near sea level, at Yale, British Columbia, and winding through British Columbia, to George Straight (not the singer, the geographical area.)

The area is rich in native story-telling, and rich and abundant in wildlife, game, and pretty much everything you would want as a wild animal living in the area, including the Frazier River, its tributaries and valleys, that feed it and keep in rich in nutrients and sediments to keep the land fertile.

OK, OK, I know I promised it wouldn't be as scientific as my Nahani Blog...lets get to what we all desire.....the Mystery.

According to Hammerson Peters excellent blog and video, on disappearances in Frazier Canyon, he tells of some of Canada's most disturbing disappearances.

In 1858, as the word of gold reached prospectors south of the border, British Columbia, governed by James Douglas, commissioned the British Army and some private contractors and laborers, to build an 18 foot wagon trail through the twisting mountain sides dubbed the Caribou Trail, due to the great number of Caribou roaming the Canyon.

Some of the first prospectors to head up the trail to seek their fortune never made it to their destination. Either through circumstances of unfamiliar travel or a new road bed taking the entire wagon trail over the cliff sides, or more macabre intervention. I dug through several paper archives, and could find no stories up to the present day, for reports of wreckage or skeleton remains in the valley. This isn't to say they are not there, or failed to make the trip (mind you, the 1800's - early 1900's papers were notorious for embellishing stories, and a simple accident is not as exciting as a vanishing without a trace). Regardless, stories and tales began of the "Phantom of the Caribou trail."

John L. Zeller, in the May 1961 article he wrote for Fate Magazine, tells of the tale of John Fillmore, a freighter who was moving supplies to the miners that had made camp for the night near Spencer Bridge. With camp set and guards in place, a strange white light was seen moving back and forth in the night sky. There was nothing to report in the morning, except they discovered three mules were missing. These mules became the first reported and became the first victims of the Caribou Phantom.

Three nights later, a similar fate found three more pack mules, to George Lateau who had made camp near Yale. Lateau sent riders on horseback ahead on the trail, and back down to the village of Yale, not a trace of the three mules was found on the trail or on the sandbars of the river below.

One of the most told stories to come out of the Frazier Valley, are the stories of Jacko, a supposed wild-man in the village of Yale. There are several stories told of just how Jacko came into the history of the area, with the most popular one being that he was found lying across the railway tracks outside of Yale. First thought to be a "drunken Native" the train crew is said to have climbed down from the engine and loaded him into a box car. Although the history gets very mixed at this point, lending doubt to the veracity of the story, it is still a great piece of Canadian folklore.

With dozens of Sasquatch reports gathered over the years, and strange lights, the Frazier Valley is a remarkable study area, full of rich history of gold mining, success and failure, and not least mystery.

Sources

- "Phantom of the Caribou trail," By John L. Zeller may 1961 Fate magazine
- "Unexplained Disappearances in the Frazier Canyon," By Hammerson Peters
- "Jack, the mysterious wild-man from Yale," By Hammerson Peter
- British Columbia government Archives

The most promising Cryptozoological candidates for success in the near future

By Chris Forbes and Raven Heather

The 21st century has already seen a handful of cryptozoological success stories. The laotian rock rat, known locally since antiquity as the kha-nyou, was formally described by western science in 2005 - and also identified as a living member of a family hitherto believed to have gone extinct, 11 million years ago (Shuker 2016). In 2006, twenty years after "an anomalous bear, looking like a cross between a grizzly and a polar bear" appeared on the first checklist of cryptozoological species (Heuvelmans 1986), the occurrence of grizzly-polar bear hybrids in nature was confirmed (Mallet 2008). In 2018, a Zanzibar leopard, previously thought extinct after an extermination campaign in the 1960s, was caught on camera (Freeman 2022), confirming prior suspicions of its survival (Walsh 2008). More recently, in November of 2022, an expedition used insight from local residents to capture video of a bird thought to have gone extinct 140 years ago (Sottile 2022).

The list goes on and on, and leads us to wonder: Who's next? Listing which cryptids are most likely to receive mainstream zoological recognition in the near future is, admittedly, not a new concept (Freeman 1999) (Coleman et al 2003). But we feel it is time to revisit this question and provide a fresh new insight into which cryptids are most likely to yield success in future cryptozoological efforts.

Allegedly extinct yet commonly sighted birds, such as the Ivory-Billed Woodpecker and South Island Kōkako are not included, as not all wildlife authorities consider them extinct. Cryptids who have an intact physical specimen in a known location, such as the Spotted Lion, Gambo, and Dobhar-chu, are not included either as their likelihood of recognition depends almost entirely on what happens to those specimens, making their chances extremely difficult to determine.

Without further ado, here are the 17 cryptids that we believe have the best chances of being undeniably proven and officially recognised/accepted before 2050.

1: British Big Cats.

The presence of breeding populations of wild big cats in the UK is indicated by 500-600 reported sightings each year (Ly 2023), one of which was made by the CFZ's own Zoological Director (Freeman 2019). In 1976 a clear cut explanation for these sightings was fully realised, the "Dangerous Wild Animals Act" or "DWAA" government policy which required special licenses for exotic pets. It is likely that rather than comply with this legal act and pay for a license, some owners released their pets into the wild (Eberhart 2002). Outside of this event, there are many sources that these animals may have originated from in history. Unlike most other cryptid types, distribution anomalies and thus BBCs are differently recognised under our philosophy. Wild cats in Britain have been largely confirmed to exist through a handful of specimens such as a puma captured in 1980 and a lynx shot in 1991, and more recently through DNA testing (Aspinall 2023). However, are these to be considered to have expanded their range? Our philosophy is that these are still merely escapees until proven to have breeding populations, thus making the offspring of these cats completely wild.

2: The Thylacine.

Since the last Thylacine in captivity died in 1936, there have been over four thousand reported sightings, many of which have been made by credible witnesses such as wildlife rangers and zoologists, and the locations of which match where surviving Thylacines are predicted to be. It would be extremely surprising if Tasmania is still officially considered a Thylacine-free island for much longer. There are also encouraging signs that Thylacines may still be present in New Guinea, where it's known for sure to have lived in the recent past and where the locals name it the Dobsenga, describing the animal with exceptional accuracy. If this marsupial is diligently sought after, at least one of these two locations will almost certainly provide undeniable proof of its survival.

3: The Orang Pendek.

While not as popular as the Yeti or Sasquatch, this is the most promising of all cryptozoological primates. If the consistent and convincing reports of prior decades were not enough, an encounter during the CFZ's 2022 expedition to Sumatra (Freeman 2022) has, in our opinion, confirmed that the Orang Pendek not only was, but is a real, living species on the island. So unless something very unexpected happens, the Orang Pendek's existence should soon be undeniably proven, perhaps by a future CFZ expedition. We agree with the hypothesis that the Orang Pendek is a member of the Orangutan genus, and we quite like the suggested binomial *Pongo martyri* first coined over a decade ago (Downes 2009).

4: Mystery Cetaceans.

From Jean René Constant Quoy and Joseph Paul Gaimard to Enrico Giglioli, many experienced naturalists over the years have observed Odontocetes and Mysticetes

Alula whale (Master Tofu)

currently unaccepted by mainstream cetologists. However, the rate at which new cetaceans have been discovered - a trend which, for the record, demolishes the popular belief that marine megafauna cannot go unnoticed in modern times - suggests that these cryptozoological whales will not have to wait much longer for their turn in the spotlight. Of particular interest are four dolphin species observed by an experienced naval captain and later described by him in an otherwise-ordinary field guide to cetaceans (Bruyns 1971). So far one of these four "mystery dolphins", the Illigan Dolphin, has been confirmed to exist, as the external appearance of the Melon-Headed Whale Peponocephala electra, which was only known from skeletal material at the time of his observations. The other three, labelled the Alula Whale, Greek Dolphin, and Senegal Dolphin, remain unconfirmed. However, truth is likely rooted in his words due to his background, and the other whales in his guide are all accurately described. While we think Cadborosaurus might also be a cetacean, and the Many-Humped Serpent probably is, we're counting them separately on this list due to their uniqueness and importance.

5: The Caspian Tiger.

A 2018 CFZ expedition to Tajikistan uncovered a substantial quantity of recent eyewitness accounts suggesting that this subspecies or population of tiger, is much less extinct than is believed by official sources. Combine this with 21st-century sightings in western China (Xu 2018), and how easily a tiger population could go relatively unnoticed in this remote and underdeveloped part of the world, and it seems likely that this will be another upcoming cryptozoological success story.

6: Giant Anacondas.

We already have a fossil record precedent for snakes reaching a maximum known length of between 13.02 and 15.58 meters (Head et al 2013). The research study which started the popular claim that this size can only be achieved in a much warmer global climate than ours, relies on the starting assumption that modern Anacondas can only grow up to 7.3 m (Head et al 2009), making its use as an argument against Giant Anacondas nothing more than circular reasoning. Anacondas live in one of the most insufficiently-explored areas of the world, they give live birth so they don't need to come onto land to lay eggs, and by staying in the water, they can rely on buoyancy to support their immense weight. Perhaps encounters with these giants aren't even as rare as they may appear to us. If an uncontacted Amazonian tribe, of which there are still several, is seeing 13-meter snakes, no-one in the western world would hear about that. And as the icing on the cake, let's not forget about the 21st-century sightings that we have heard about, like that of a 12-meter individual in Guyana a few years ago.

7: Giant Crocodiles.

As with the Giant Anaconda, we have a prehistoric precedent for Crocodilians reaching immense sizes, the largest known fossil species measuring well over 12 meters (Aureliano et al 2015). We would not be surprised if a Nile or Saltwater Crocodile exceeding 9 meters sets a new official maximum size for the genus. Both species are doing quite well, listed as 'Least Concern' by the ICUN, so we predict that exceptionally-large individuals will continue to grow into existence every now and then for the foreseeable future, making the recognition of a giant specimen almost inevitable (this same point can also be made in favour of the aforementioned Giant Anaconda).

8: The Yeti.

Recent decades have seen compelling indications that this infamous species of Great Ape still lives in mainland Asia. In China, hair samples of the Yeren (which is likely the same thing as the Yeti) have been tested with promising results. Scientists at Fudan University found, and western scientists have independently verified, that the iron-zinc ratio in hair samples is unlike that of human or monkey hair. At East China Normal University, biologists used a scanning electron microscope to examine hair samples and similarly came to the conclusion that they did not come from any recognised primate. And over in Bhutan just a few short years ago, environmental DNA was tested and found to contain DNA belonging to an unrecognized species of Great Ape. Now the Yeti's biggest remaining mystery is, who will be the first to bring back a specimen?

We suspect it will most likely be Chinese cryptozoologists who make the breakthrough, as the Yeti/Yeren lives hardly a few days' drive from their laboratories, making fieldwork easier for them.

9: Cadborosaurus.

Cascadia remains the top place to look for sea serpents, and judging from the number of 21st-century sightings, Caddy is very much alive. The fact that some of this cryptid's witnesses have actually been scientists reinforces the case for its existence too. An intriguing possibility is that, if previous research on Caddy's reproduction is accurate - that Cadborosaurs use Saanich Inlet from mid-June to the end of July as a breeding site, and that they give birth at night on deserted beaches or in shallow water - then these locations could be observed at this time using sonar, camera drones, or some other technique. It is worth noting that some, not all but some, of the Caddy sightings of the past few years may be misidentifications of a swimming moose, a group of swimming meese, or the next entry on this list.

10: The Many-Humped or "Classic" Sea Serpent.

With its near-global distribution among the world's oceans, it can't be much longer before one is caught on camera at close range and in exceptional detail, collides with a ship resulting in a type specimen being acquired, is caught by unsuspecting fishermen, or is found in the form of a washed-up carcass. If undeniable proof of this extraordinary cetacean's existence does not come from the open ocean, it could easily come from one of the freshwater bodies it has been observed in - probably Lake Okanagan, which we believe is the most probable of all supposedly monster-inhabited lakes to have cryptozoological residents in the present day, or the nearby Lake Manitoba. Environmental DNA testing could greatly assist the search in these lakes. As for what exactly this "serpent" is, we consider the Ziphiidae and Basilosauridae to be the two strongest contenders for its taxonomic identity.

11: The Mokele-mbembe.

The quantity and quality of this reptile's sightings - by locals, westerners, and non-western outsiders - indicate that an immense long-necked enigma continues to live in Africa's most remote rainforest. Unfortunately, the extreme cost and difficulty of conducting fieldwork in the least charted and deepest jungle in Africa has thus far prevented the acquisition of physical proof. However, this same extreme isolation may also protect the Mokele-mbembe from the damage industrial civilisation is causing to less remote environments, allowing the species to persist for decades into the future.

12: The Mapinguari.

The quantity of native Amazonian sightings of what is clearly a living ground sloth - probably a Megalonychid (Oren 2001) - leave little room for doubt. Like the Giant Anaconda, the Mapinguari inhabits remote areas of Earth's largest rainforest, where almost anything would be capable of hiding from all but the occasional eyewitness. If searched for at the right time of year, in the right part of Brazil (Serra do Divisor in Acre and the Karitiana tribe reservation in Rondônia are both promising sites), and in a manner that takes into account it's foul scent, which has thwarted previous capture attempts, it seems like one of the most likely cryptids to crawl into mainstream textbooks in the near future. Even mainstream zoologists are beginning to take the possibility of the sloth's existence seriously (Wilson et al 2022).

13: The Almasty.

One of the greatest cryptozoological breakthroughs of the 21st century so far was the identification, albeit not official recognition, of Abkhazia's Almasty as a unique species within the genus Homo thanks to Bryan Sykes' DNA analysis (Sykes 2015). Unfortunately, very shortly after Sykes' passing in December of 2020, attempts began almost immediately to discredit his scientific research and findings (Margaryan et al 2021). However, any doubts we may have had about the reality of the Almasty can be put to rest by the CFZ's own encounter with one in Russia in 2008. And if the Almasty population is truly on the increase, then only bad luck, the political situation within its habitat range, and the natural tendency for people to assume our uniqueness in the animal kingdom will stand in the way of its official recognition. In honor of the geneticist who found this cryptid's place in the Hominid family tree, we suggest granting the Almasty the binomial *Homo sykesi*.

14: Giant Monitor Lizards.

There are a few different ways that the Komodo Dragon could be dethroned from its title as the largest living member of Varanus. In Papua New Guinea, sightings have continued into the 21st century of individuals of the Salvadori Dragon *Varanus salvadorii* growing well past its "official" maximum size. On the Indian subcontinent, sightings of a giant monitor locally known as the Buru or Jhoor have continued well past its alleged extinction date of the late 1940s. In Australia, the survival of V. priscus or something similar is indicated by many modern sightings - one of which was made by a professional herpetologist, at close range, with a clear frame of reference for estimating the lizard's size. Unrecognized species of giant monitor may also be the identities of East Asia's Long and Central Africa's Nguma-Monene. One way or another, it's very likely a record-breaking monitor will be caught. The Mokele-mbembe may also be a member of the Varanidae family, but probably not of the Varanus genus itself.

15: The Lusca.

While attempts have been made to discredit the identity of the St Augustine carcass (Pierce et al 1995), the evidence that a species of colossal octopus lives in the Caribbean, and especially the Bahamas, has been piling up for over a century since that infamous incident. Even biologists who typically take a hardline stance against cryptozoological explanations for aquatic monsters, are starting to admit that a giant octopus is the best explanation for recent sightings and attacks, and could remain hidden in the Blue Holes of Andros and elsewhere (Wade 2016). Remotely-operated vehicles, and perhaps even small manned submersibles, could play a large role in exploring the labyrinth of extensive undersea caves that this species inhabits.

16: The Long-Necked Sea Serpent.

Known more often by various local names, the Long-Necked Sea Serpent is perhaps the most famous cryptid of all, with more written about it than any other. This is unsurprising, especially since it appears to have the most extensive total habitat range of any cryptozoological species. While there hasn't been much exceptionally promising news from Loch Ness for a few years now, sonar readings in late 2022 suggest that the species is still present in Lake Champlain (Elizabeth 2022). The Longneck remains the most actively searched-for aquatic cryptid, and it seems like it's truly just a matter of time before undeniable proof is acquired, either at one of the numerous lakes it's been reported from, or out at sea.

17: Giant Eels.

While doubt has recently been cast on the plausibility of giant *Anguilla anguilla* in Loch Ness (Foxon 2023), many modern aquatic monster reports do seem to refer to large eels - both giant members of recognised species, and members of unrecognized species. In particular we suspect that eventually Heuvelmans' Super-Eel will swim in front of an ROV or oil rig camera. While it is not seen as often as the mammalian sea serpents, this is explained by the fact that, like most fish, the Super-Eel does not need to come to the surface to breathe.

Honorable mentions: In addition to the 17 cryptids listed above, another 10 seem somewhat promising but didn't quite make the cut. These are the Allghoi Khorkhoi, Crested Crowing Cobra, Ebu Gogo, Emela-ntouka, Gul, Japanese Wolf, Kawekaweau, Sasquatch, Steller's Sea Cow, and Trinity Alps Salamander. Any of these could become major cryptozoological success stories in the near future, but we wouldn't advise placing bets on them. Either way, we predict that the years ahead will be an exciting time for cryptozoology - provided that the cryptozoological community remains optimistic and enthusiastic, and that the loudest voices in opposition to the emerging discipline are not permitted to ruin everything.

Sources cited:

- Aspinall A. 2023. DNA test is 'definitive PROOF' huge black panthers are prowling UK countryside. Mirror. Available at https://www.mirror.co.uk/news/weird-news/definitive-proof-huge-black-panthers-29951838.
- Aureliano T, Ghilardi AM, Guilherme E, Souza-Filho JP, Cavalcanti M, Riff D. 2015. Morphometry, Bite-Force, and Paleobiology of the Late Miocene Caiman Purussaurus brasiliensis. PLoS ONE 10(2):e0117944.
- Bruyns WFJM. 1971. Field guide of whales and dolphins. Amsterdam (NL): Tor Books.
- Coleman L, Huyghe P. 2003. The field guide to lake monsters, sea serpents and

other mystery denizens of the deep. New York (NY): Tarcher.

- Downes J. 2009. Expedition Press Release. CFZ Sumatra 2009. Available at https://cfzsumatra09.blogspot.com/2009/09/expedition-press-release_29.html.
- Eberhart GM. 2002. Mysterious Creatures: A Guide to Cryptozoology. Santa Barbara (CA): ABC-CLIO.
- Elizabeth K. 2022. Our Evidence. Katy Elizabeth. Available at https://www.katyelizabeth.org/our-evidence.
- Foxon F. 2023. If it's real, could it be an eel? [Preprint].
- Freeman R. 1999. Monsters of the next millennium. Animals & Men. 20:13.
- Freeman R. 2019. Adventures in Cryptozoology: Hunting for Yetis, Mongolian Deathworms, and Other Not-So-Mythical Monsters: Volume I. Coral Gables (FL): Mango Publishing Group.
- Freeman R. 2022. In Search of Real Monsters: Adventures in Cryptozoology Volume 2. Coral Gables (FL): Mango Publishing Group.
- Freeman R. 2022. Sumatra 2022. CFZ. Available at https://cfz.org.uk/sumatra-2022/.
- Head JJ, Bloch JI, Hastings AK, Bourque JR, Cadena EA, Herrera FA, Polly PD, Jaramillo CA. 2009. Giant boid snake from the Palaeocene neotropics reveals hotter past equatorial temperatures. Nature. 457:715-717.
- Heuvelmans B. 1986. Annotated checklist of apparently unknown animals with which cryptozoology is concerned. Cryptozoology. 5:1–26.
- Ly C. 2023. Are big cats like black panthers and leopards really roaming the UK? New Scientist. Available at https://www.newscientist.com/article/2378197-are-big-cats-like-black-panthers-and-leopards-really-roaming-the-uk/.
- Mallet J. 2008. Hybridization, ecological races and the nature of species: Empirical evidence for the ease of speciation. Philosophical Transactions of the Royal Society B: Biological Sciences. 363:2971–2986.
- Margaryan A, Sinding MS, Carøe C, Yamshchikov V, Burtsev I, Gilbert MTP. 2021. The genomic origin of Zana of Abkhazia. Advanced Genetics. 2(2).
- Oren DC. 2001. Does the Endangered Xenarthran Fauna of Amazonia include remnant ground sloths? Edentata. 4:2–5.
- Pierce S, Smith G, Maguel T, Clark E. 1995. On the Giant Octopus (Octopus giganteus) and the Bermuda Blob: Homage to A. E. Verrill. The Biological Bulletin. 188(2):219-230.
- Shuker KPN. 2016. Still in search of prehistoric survivors. Greenville (Ohio): Coachwhip Publications.
- Sottile Z. 2022. This bird hadn't been documented by scientists since 1882. Then they captured video of it in Papua New Guinea. CNN. Available at https://edition.cnn.com/2022/11/19/world/black-naped-pheasant-pigeon-scn-trnd/index.html.
- Sykes B. 2015. The Nature of the Beast. London (UK): Coronet.
- Wade J. 2016. Terror in Paradise. River Monsters. Bristol, England, United Kingdom: Icon Films.

- Walsh MT, Goldman HV. 2008. Updating the inventory of Zanzibar leopard specimens. CAT News. 49:4–6
- Wilson OE, Fortelius M, Saarinen J. 2022. Species discovery and dental ecometrics: Good news, Bad News and recommendations for the future. Historical Biology. 35(5):678–692.
- Xu DC. 2018. Mystery Creatures of China: The Complete Cryptozoological Guide. Greenville (Ohio): Coachwhip Publications.

Highway 413: Such a bad Idea.
David Scott

The Ontario Government has recently paused its 'guns blazing' stand on the building of a new highway from Hwy 400, through Caledon in the greenbelt area, west of the greater Toronto Area.

The Reasoning for building it was decreased commute time and quicker delivery times. That time? 30 minutes. OK, I am all for a solid supply chain, commute time not so much, as it is a choice where you live. Perhaps stop grabbing farm land to build mammoth sized homes, and build on existing blighted area with more affordable housing, this may address those problems.

The cost to the environment is staggering. I do not confess to having trouble keeping a cool head when I was gathering facts, fact checking, and doing research for this essay. I do not know who ok'd this to move forward, they either didn't do the work, didn't care, or in my opinion, were slipped a few bucks (my opinion only and does not in anyway shape or form represent the CFZ).

First off, 127 water shed crossings, adding garbage and carbon pollution to the waterways. It crosses Indigenous land, which doesn't belong to us, including and not the least important ceremonial lands, and burial areas that have been there, long before the GTA even existed. Considering how truth and reconciliation are spouted by the government, this just shows the lack of respect and true concern, our First Nation Peoples are truly being shown.

Agricultural Farmland is being paved over, as well as hiking trails, walking paths and green spaces. In a time when Ontarians are adapting a more healthy life style and an increased population needing more food, is this a good idea? They claim to be creating jobs, I know of quite a few job creations that can be formed, existing transportation structure repairs, POTHOLE repairs, access to remote areas now only accessible by old cart trails, so the improvements to these areas would have far less environmental impact.

Now the killing begins.
The world is amidst a rapid extinction of a staggering 55,000-73,000 species every year. 150-200 plant species go extinct everyday! (Statistics compiled by Kristen Holder for AZ Animals website)

In the green belt area of the proposed highway, we have the following critical and endangered species and plant life. Negatively affecting 29 species listed under the Federal Species at risk act (see citation link bottom of essay).

In a report by Biologist Karl Heide, and Dr. Ryan Norris, of the University of Guelph, they identified 12 Birds, 9 Amphibians and reptiles as well as 8 Fish, insects and trees.

- Blandings Turtle-highway dependent on water quality to survive.
- Butternut tree - few butternut trees where unaffected by the butternut canker, a fungal disease to a point where only a few remain totally disease free, many of them exist in the greenbelt area on the proposed highway.
- Eastern Meadowlark- More than half the proposed area of Highway 413 is the breeding ground for the Eastern Meadowlark who are already in

steady decline due to urbanization.

- Eastern Ribbon snake- This semi aquatic snake is threatened by loss of wetland and road mortality. Proposed connection to highway 427 for 413, would add significant pressure for the snake that lives in this small area.
- Jefferson Salamander and uni-sexual Ambystoma - Suffering from one of the highest mortality rates already, the building of highway 413 and the connectors to highway 401 and highway 407, would not only contribute to pollution in the ponds they are known to live and breed in, but also block migration route access.
- Least Bitten- these small low flying birds already suffer from auto collisions and breed in the proposed area, including the proposed land taken back from the Heart lake Conservation area. Fragmenting wetlands will have a very strong impact on the survival of the species.
- Monarch Butterfly- If nothing else will stir a persons wonder, then the annual migration of monarch butterflies will, like in the 1970's when I was a child. If you didn't see these iconic butterfly on your walk to school, you had a bad day. I hike and spend much time in our yard, and in

green spaces and forests. I can count on one hand the number I have seen in the 2023 summer season. Pesticide use and loss of milkweed plants have resulted in a devastating decline of its population, and studies have proven, road traffic has increased with volume.

- Rapids Clubtail - This Dragonfly is listed as endangered and extremely sensitive to degradation of river conditions. They live in only 4 rivers in Canada, two of which are the Humber River and Credit river, and are both affected by the proposed route of highway 413.
- Red-Headed Woodpecker - This Endangered species of Ontarian Woodpeckers, is extremely scattered, but is core area sits right in the middle of the proposed area for the highway that would destroy breeding habitat. And could possibly lead to its extinction.
- Redside Dace – Listed as one of most endangered types of fish species in

Canada. Primary impacts for this fish are pollution to streams and waterways. They would become even more threatened by the loss of habitat under the path of the highway and additional pollution.

- Rusty-Patched Bumble Bee – At risk of global extinction! Climate effects in the area would ensure loss of habitat and increasing the likely hood of eradication of this Bumble Bee.
- Short-Eared Owl – Suffering from one of the steepest declines of any bird

listed by the Federal Species act. It has lost 90% of its population since 1966. They are known to exist at the east side of the 413, at the connection area for highway 400. Building in this area would eliminate them from this area.

- Wood Thrush – a relative of the robin, the Wood Thrush live in almost

50% of the forest fragments slated to be removed by the highway 413.

- Western Chorus Frog – Who hasn't heard the throaty croak in spring of several Chorus frogs? Its a staple of every outdoor movie, TV show and a sign of spring to many Canadians. I know one person who has heard none near his home the past few years, and was akin of the end of winter. I live in one the fastest growing cities in Canada, and urban sprawl has destroyed so many habitats and polluted the local waterways, to a point I have to drive far north, to regain a true feel of Nature.

Frogs breath through their skin, so the additional run off of road salt from highways, and the additional pollution, would greatly impact the sustainability of the species, Don't be like London, save what you have and work towards better mass transit and rail to alleviate congestion, your 30 minutes is not worth the eradication of a critical environment corridor.

David A. Scott
Canadian Head
of the Centre of Fortean Zoology.

The monster of Lake Atter, Austria

Ulrich Magin

Several years ago, I published a study of the first year of the Loch Ness Monster, first in the journal *Bipedia* and then as a chapter in my book *Investigating The Impossible* (Anomalist Books 2011). I found that the first observations of "the monster" had little in common, until isolated and more sensational reports formed the image that is with us still today.

This situation is almost identical in the first phases of any lake monster that becomes known – reports vary, and their only common denominator is that something large and unusual was seen. Only later, a stereotypical monster is established, and this phantom portrait then serves as a model for future witnesses. Loch Ness was lucky, because first the London press, then the American press, noticed the local reports and spread them from there, to the world.

The whole thing could have happened as early as 1902 at Lake Atter (Attersee) in Austria, where exactly the same thing happened, a sort of (however, largely unnoticed) blueprint for what happened 30 years later in Scotland.

Lake Atter

The Attersee is located in the Salzkammergut cultural region, in the Vöcklabruck district of Upper Austria. The River Ager, which is an outflow of the lake, runs into the River Traun, and thus into the Danube. The lake, a typical post-glaciation formation, is situated 469m (1,539 ft.) above sea level, with an average depth of 85m (279 ft.) and a maximum depth of 169m (554 ft.) – making it the third deepest lake in Austria. Because of its height above the sea, it is very unlikely that any large sea animal could enter it (unlike Loch Ness, which is regularly visited by seals and possibly even small whales). The lake is around 18 km long and three and a half km wide, and has a water surface of 46 km². It has been inhabited since the Neolithic and is now considered a top tourist destination in Austria, with tourist numbers steadily increasing. Around 1932, some 112,750 overnight stays were recorded, 1980 around 795,000 and in 2019 more than 8.5 million. If a large unknown creature lived in the lake, it couldn't go unnoticed for long.

The wave of 1902
We will look at what happened in 1902, at Lake Atter, through the lens of the local press of the time [comments in square brackets are mine]. The first report that I could discover comes from the "Grazer Volksblatt" of August 26, 1902. On page 8 we read a cautiously worded and sceptical account:

> ** A mysterious phenomenon at Lake Atter.*
> *From Weyregg on Lake Atter a letter was addressed to the 'N. Wr. Tagblatt' [Neues Wiener Tagblatt] [by Dr. Gustav von Wunschheim, Vienna]: "Allow me to tell you about a very remarkable natural event, in the hope that, by distributing it in your widely read paper, an explanation of it might be found. Today [22 August] at a quarter to twelve o'clock, from the balcony of my apartment which offers an unobstructed view of a large part of Lake Atter, I noticed an object behind the headland of Weyregg in a distance of about one and a half kilometres which shot along the surface in an incredible speed and in a straight line in a southward direction. Quickly grabbing our glasses, we followed this utterly mysterious apparition, which was still moving across the lake in front of us with undiminished speed, and we could clearly distinguish that the apparition was about four to five meters long [16 ft.] and towered about a foot above the water. It had two humps which kept their distance and were about half a meter high [2 ft.], and which appeared to be fins. The apparition disappeared in the region between Nußdorf and Parschell, but left a wake that could be followed for a quarter of an hour as a dark dead-straight line in the slightly choppy water. According to the map, the distance covered was about three miles, and the time it took the apparition to negotiate this distance was one and a half to two minutes at the most. So the speed was at least two kilometres per minute.*
>
> *At first we thought of a man-made object which had been*

launched in the lake for unknown experimental purposes, but we soon realized that even if it had been a torpedo it would never have managed such an incredible speed in such a large dead-straight line. The only explanation we could think of was that a fish, enormous for an inland water, for reasons unknown, made this fast journey across the surface of the water.

It would be very desirable that this phenomenon were also observed by others and perhaps from a much closer distance, and possibly these lines will prompt relevant communication to your esteemed newspaper." –: Well then! The famous sea serpent has finally appeared, and we open the silly season.

The original newspaper report, in the "Neues Wiener Tagblatt" from 23 August 1902, p. 25, adds nothing to the reprint. Soon after, a second unusual observation appeared in the local press. The "Linzer Volksblatt" reported on August 29, 1902, on page 5, on two additional observations, one well before the current run:

A colossal fish in Lake Atter. We have received this communication from Lake Atter: When the morning steamer leaving Unterach had passed Parschall [today: Nussdorf-Parschall] and was at the hills near Zell last Sunday [24. August 1902], the sub-captain Ludwig Souvent, who was busy in the cabin, noticed a strong bloch?! [this ?! is in the original, it is an unknown word, and the contemporary editor added the?!, apparently it means a floating tree trunk] at a considerable distance from the boat. Since he feared a collision if he continued on his course, he attracted the helmsman's attention by shouting and signalling to avoid it. Curious, the above-mentioned bent over the ship's hull and soon saw not a wooden colossus, but a fish monster, grey-blue, its back sticking out of the water more than half a man's height and on it a fin of a good half meter [2 ft.]. When the monster, on which he could detect neither head nor tail, got closer to the whirling paddle wheel, it immediately disappeared into the depths. Several years ago there was some talk on how the Stockwinkl mill boy, on his way home from Steinbach, encountered just such a colossus of a fish. At the time, his description was thought to be a hoax, which earned him the nickname of Jonas of Stockwinkl. (editorial note: A few days ago in the 'N. Wr. Tagblatt' a Mr. von Wunschheim, who had noticed the big fish from his villa, described the appearance of it in such a drastic way that many took the fish for a – sea serpent.) [i.e., they thought it was an invented tale.]

The identical report was printed in the "Salzkammergut-Zeitung" on August 31, 1902 (p. 5). So not only was a second, independent sighting of the monster reported, but another observation of the "colossal fish" – "several years ago" – was now recalled. At that time, because the account was solitary, it met not belief, but derision.

But the new reports were also mocked. The "Freie Stimmen" of August 30, 1902 could not resist calling the report "Sea Serpent", the then current term for a silly-season hoax:

In keeping with the dog days, all sorts of strange apparitions appear in the newspapers, as they do every year, the oldest man, the angry dog, the child-eating pig and the shark from Fiume [today Rijeka in Croatia]; the latter, five meters long, has actually been captured this time. A Viennese paper even spotted an eerie big fish in the Attersee, an offshoot of the good old sea serpent, and published an editorial about the same.

The "Grazer Volksblatt" also briefly mocked the monster on August 13, 1902 (page 8) and listed it together with other irrelevant summer topics which had caused some excitement (the "Sea Eyes" referred to were lakes on the frontier between Austria and Galicia, which had led to border disputes):

** This week. [...] The silly summer season is already coming to an end, and in addition to the monkey theatre [meaning: much ado about nothing], the two temples of the muses have opened again. This year the well-known sea snake with the obligatory hump was seen 'in Lake Atter' just once and the question of its existence was pushed into the background by the question 'Whom do the Sea Eyes belong to', which must be so important for the whole world.

A few days later, the event was recapitulated, again in the "Salzburger Chronik für Stadt und Land" on September 2, 1902:

*Strobl, September 1st. (Brief news from the Salzkammergut.)
On several occasions already, a fish monster of four to five*

meters length has been repeatedly noticed in Lake Atter. The speed of this fish colossus was at least two kilometres per minute. At first the sub-captain Souvent thought he saw a strong bloch some distance from the steamboat. This apparition caused a great stir in the area and was subjected to much attention.

Explanations were offered. Contemporary German newspapers often had – as a kind of pre-Google Google – a column that answered inquiries by readers about the facts of a topic. So when an anonymous reader wanted to know what the thing in Lake Atter was, the "Salzkammergut-Zeitung" published the following item on 5 October 1902, on p. 6:

Sea monster. Regarding your inquiry, we specifically contacted an expert friend of ours on Lake Atter, and he replied with the following about this mysterious sea monster, which has been repeatedly observed in the waters of Lake Atter recently: "That this fish could be a so-called 'sea monster' is, in my opinion at least, an exaggeration. There are fish of abnormal size here (salmon, pike), but such giant fish as this 'sea monster' is supposed to be simply do not occur in our waters. I suppose that the monster which was observed is not one but several fish, for example two or three very large salmon. At spawning time, they often swim across the lake one after the other at a lightning speed, so that to the surprised observer they appear as one single fish. Two or three such large fish can easily add up to a length of three to four meters. In any case, there are also ambush predator fish which for reasons unknown have left their ambush sites. It must also be borne in kind that the lake can be very deceptive. During my three-week stay in Vienna, on the occasion of the international fisheries exhibition, I spoke to various experts on this matter; they all agree with my view."

This same short piece can also be found in the "Linzer Volksblatt" of 7 October, 1902. A last encounter reached the press in December, 1902. On 11 December, the "Tages-Post" of Linz had this to say (on p. 5):

(The Lake Atter Sea Serpent.) We received a letter from Weyregg, 8th of this month: This summer, a lake monster was repeatedly observed in Lake Atter, appearing here and there

and frightening the bathers. This monster was seen again very recently. On the afternoon of the 6th of this month the fire chief Fr. Kestler, standing on the shore of the lake at Seewalchen, noticed a black stripe on the water surface about 50 m away, which he initially thought was a block. But he soon realized that it had to be "Jonas", as it is jokingly called, and hurried to his home to get a rifle and greet it with a precise bullet. When he returned, however, every trace of this mysterious fish had disappeared. When will somebody manage to get hold of it?

And then the reports in the papers died. A Nessie phenomenon could have developed from these observations, and in 1933 Loch Ness would have been called the "Scottish Lake Atter". But it didn't happen, because the world press didn't react to the reports. As far as I could ascertain, the news did not even reach Germany. I could also not trace a single reference in Papers Past with all New Zealand newspapers, or in the Australian Trove. And there were only three more press articles about the incident in the Austrian digital newspaper library, and no single sighting.

Life after 1902
Reporting on a sighting of the tatzelwurm nearby, the "Tages-Post" on Linz briefly noted, on 17 July, 1904, p. 6, that the sea serpent was "no longer an unusual animal for us, since one was noticed in Lake Atter not too long ago. Now, even scholars may delve into the question of the existence of the sea serpent – for us this monster does not even remain a riddle – it is just a fantastic fairy tale. And as the water has its sea serpent, the land has its legendary mountain serpent."

Twelve years later, the Lake Atter serpent was resurrected. On 12 April 1914, the "Salzkammergut-Zeitung" (p. 13) printed a satire in local dialect claiming a hillbilly had killed the Lake Atter sea serpent.

The 1914 fake photo.
This referred to an April Fool's hoax (even including a photograph of the creature) that had been in the newspapers edition of March 29 (the last before 1 April). We read in the "Salzkammergut-Zeitung" on 29 March, 1914, p. 1 of the supplement:

Salzkammergut-Zeitung.

Nr. 9,   Gmunden,   29. März 1914.

Tirana, die Krönungsstadt Albaniens.

Tirana, die ehemalige Hauptstadt Albaniens, liegt etwa acht Stunden von der Küste entfernt. Sie ist kulturhistorisch, wie landschaftlich die schönste und interessanteste Stadt Albaniens. Unser Bild zeigt den Haupt- und Marktplatz von Tirana während eines Wochenmarktes, auf dem die interessantesten Typen der Bewohner Albaniens versammelt sind.

Tötung des Seeungeheuers im Attersee.

Seit einer Reihe von Jahren kamen Nachrichten aus Weyregg, Steinbach, Unterach, Nußdorf und Kammer von einem schrecklichen Ungeheuer, das sich in den Gewässern des Attersees aufhält. So sehr man sich sich auch bemühte, desselben habhaft zu werden, war dennoch bisher alle Arbeit vergeblich. Es war dies umso mehr zu bedauern, als die zahlreichen Sommergäste an den Ufern des Attersees sich schon nicht mehr getrauten, im Attersee zu baden, aus Angst, von dem Ungeheuer erwischt

Die Tötung des See-Ungeheuers im Attersee.

und aufgefressen zu werden. Sogar die sonst doch so kühnen Ruderer und Segler erfaßte ein gewisses Zittern und Zagen, wenn sie die Fluten des Attersees mit ihren Schiffen durchqueren, in der gewiß nicht unbegründeten Furcht, von dem geheimnisvollen und jedenfalls auch blutdürstigen Ungeheuer plötzlich angegriffen zu werden. Tatsächlich geschah es nun dieser Tage, daß ein Holzknecht, vulgo „Federn Sepp“, von dem heißhungrigen Tier, als er eben mit der Hacke sich zur Wehre setzen wollte, aus dem Boote in den See geschleudert wurde. Im Sturze kam er aber gerade auf den Rücken des Fisches zu

The Lake Atter Sea Monster Killed

For a number of years, news has been coming from Weyregg, Steinbach, Unterach, Nußdorf and Kammer about a terrible monster residing in the waters of Lake Atter. No matter how much effort was made to get hold of it, all this work has been in vain so far. This was all the more regrettable as the numerous summer guests on the shores of Lake Atter no longer dared to swim in Lake Atter for fear of being caught and devoured by the monster. Even the otherwise bold rowers and sailors felt a certain trembling and hesitation when they crossed the waters of Lake Atter with their boats, in the certainly not unfounded fear of being suddenly attacked by the mysterious and bloodthirsty monster. It actually happened recently that a woodworker, vulgo "Pedern Sepp" [the Austrian rendition of a personal name, possible Peter Josef], was thrown out of his boat and into the lake by the ravenous animal, just as he was about to defend himself with his hoe. But in his fall he came to sit on the back of the fish and managed to smash its skull with his hoe, as shown in the picture above. Next Wednesday there will be a general fish dinner in Mr. Fürthauer's inn in Steinbach at Lake Atter, to which the residents of Lake Atter are most politely invited. The sea monster killed so bravely by Pedern Seep will be boiled and then eaten. During the festival / eaten, it will be decorated with the golden medal of merit by Pedern Sepp so bravely / Weißenbach as a reward for his daring deed. So come you all to Steinbach next Wednesday.

In the last paragraph, several lines are missing (marked by a /), either to make the story more interesting or because of carelessness by the printer. Is the newspaper's claim that there had been sighting reports for "a number of years" part of the hoax, or is it true? I have not encountered any reports between 1902 and this April Fool's article, and still – why resurrect something forgotten after 12 years? However, if there were any, I have not been able to track them. Perhaps, with more newspapers available in digital form, this monster story can be expanded.

Hints at earlier sightings

There are some hints that something like a monster may have been spotted in earlier times. On June 1, 1930, the "Salzkammergut-Zeitung" reported on local legends and tales. Among these, the following folk story appears on page 4:

The sea monster.

According to ancient tradition, there is a monster of almost unbelievable size in the depths of Lake Atter. Usually it lies motionless at the bottom of the lake, but sometimes it stirs a little, then the lake foams up wildly and throws up big waves, without even a whispering wind being noticeable. (P. Amand Baumgarten: Aus der volksmäßigen Ueberlieferung der Heimat, p. 35 [1864]) The sea monster comes to the surface every twenty to thirty years. The last one to tell that he saw the giant fish was the old postman Thomas, and it protruded a good twenty meters [70 ft.] out of the lake. The fishermen then have to be careful and draw in their large nets in good time. Today, of course, the giant nets have disappeared. But years ago, the Morganhof, the Hiaslbauer in Buchberg and the

The ship pulpit with the monster at Fischlham church. (photo: Isi-wal, wikimedia.org)

owner of the castle in Burgau, which has sunk into the lake, fished with huge nets that have long since become obsolete because of their cost. These nets were trawl nets, and a trawl fisherman is distinguished from the simple angler. (Erwin Volkmann: Die deutsche Stadt im Spiegel alter Gassennamen, S. 85 [1926])

We will encounter these "waves without wind" again. They were also reported at Loch Ness and Loch Lomond in the 18th century.

Another item is the "ship pulpit" (a pulpit in form of a boat, as fits a fishing community) of the 1447 gothic style church of Fischlham. It shows a sea monster, but certainly not as a depiction of a local, rarely seen creature, but as a reference to the devil. As the "Salzkammergut-Zeitung" says on 1 September 1949, on p. 10:

The showpiece of the church is the magnificent baroque pulpit, a ship pulpit, the work of the two Lambach artists Adam Racher and Franz Xav. Leitner.

This magnificent work of a pulpit illustrates the lovely scene of the sermon at the Sea of Genezareth and the miraculous fishing, it is very artistically and charmingly represented. In the boat of the pulpit we see a life-size sculpture of Christ, Peter and James, with God the Father above them, a group of angels; under the hull, a sea monster symbolizes the evil enemy. The Latin inscription, borne by an angel, translates as: "But at your word I want to throw out the net."

There are several such pulpits in the general region, another at Gaspoltshofen and a third at Traunkirchen. All three are some 20 miles from lake Atter.

Waves without wind
Clearly, the different observations of the monster (as a floating tree trunk, a giant fish, and a hump with fins) are insufficient to form a clear image of the creature. This was also true of Loch Ness in the summer of 1933, where we do indeed find similarly divergent descriptions (floating log, boat with keel up, crocodile, one to twenty humps, and giant fish). Apparently, eyewitnesses have a tendency to attribute disparate phenomena to a single cause, and they often refer to the folktale of a giant fish. What triggered the individual observations is difficult to define in retrospect. But if a monster lives in Lake Atter, especially

one so impressive, it should have been spotted again – if not in earlier times, then at least after 1902. If we assume a visitor from the sea, the obstacles are almost beyond imagination.

The reports about a drifting log may just have been encounters with a tree trunk. However, the observation from Weyregg, with which the episode began, as well as the folktale about the giant fish, may find a different explanation. In Lake Atter, as in many lakes of comparable size, there are internal seiche waves which can sometimes reach the surface. This phenomenon manifested itself dramatically on July 24, 1930 between 4p.m. and 6p.m. At that time, the "Kleine Volks-Zeitung" reported on July 26, 1930 on page 6:

> *Tides in Lake Atter.*
> *The water receded by more than ten meters. – A strange natural phenomenon.*
>
> *According to reports from Schörfling, 25th of this month, the water of the Lake Atter suddenly receded more than ten meters in the shallow areas on Thursday between 4.30 and 5 p.m. The water of the River Ager, which drains from the lake, suddenly pushed back into the lake, and it seemed as if the Ager was emptying into the lake. The boats that were in the Ager ran aground. A small Trauner [type of boat] going down the river had to make the greatest effort not to be pushed back into the lake by the water suddenly flowing up river. Yesterday, too, the lake level ebbed away at certain intervals in the shallow areas and then flooded back.*
>
> *The whole process gives the impression of a sudden ebb and flow on the Lake Atter; this phenomenon has only been observed in the Kammer-Seewalchen area; it has not been observed in Unterach until now. Fresh news say that the fluctuation lasted about a quarter of an hour the day before yesterday; yesterday the oscillation height of the water level was already much lower. The hydrographic department of the state reports that these are apparently so-called "standing waves".*
>
> **"Standing Waves."**
> *Report by the Chief building councillor Dr. Rosenauer, Hydrographic State Department, Linz.*

Chief building councillor Rosenauer from the Hydrographic State Department in Linz gave the following information to one of our employees about the phenomenon that laypeople have associated with the Italian earthquake:

These so-called "standing waves" were first observed in Lake Geneva and were studied in detail by Professor Forel. There they are 1 to 1.70 meters [3 to 6 ft.] high. This phenomenon is caused by a rapid change in air pressure or by wind gusts pressing down on the lake surface, but not by earthquakes. If one thinks of the surface of the lake as something like a seesaw and imagines that a strong pressure is suddenly exerted on one side of this board, while no change takes place on the other side, then one has approximately the idea of what science knowns as a "standing wave". Due to the effect of the air pressure, the level of the lake, which is in equilibrium, is thrown out of its position and then swings up and down like a seesaw, until after a certain time, often after days, the calm returns. Even an apparently quiet lake surface is constantly in such a rocking motion, similar to water in a shallow pool, which, even when the pool appears quiet, is always swinging back and forth.

In general, however, these lake oscillations are so slight that they are hardly noticed. The duration of the oscillations is very specific for each lake, like a pendulum, and can be calculated exactly. At Lake Atter it is around 21 minutes, but since oscillation nodes can also form, a multiple of this oscillation period is also possible. Of course, transverse oscillations can also occur, but they have a different period of oscillation. This is three and a half hours at the Attersee. Connections of longitudinal and transverse layers, if they occur, are very difficult to calculate. The height of such a lake level fluctuation is usually greatest at the ends. So far, it has hardly been more than ten centimetres at Lake Attersee. On Thursday, however, an exceptionally high oscillation was observed, the extent of which has not yet been precisely determined, but which should not have exceeded eight to thirty centimetres [3 in to 1 ft.] in total. On gently sloping shores, this lowering of the level can easily be felt in such a way that

Lake Atter, from the air
https://en.wikipedia.org/wiki/Attersee_(lake)#/media/
File:Aerial_image_of_the_Attersee_(view_from_the_southwest).jpg
(photo: Carsten Steger, wikimedia.org)

the water suddenly recedes about a meter. The period of oscillation of Thursday's phenomenon was the one known for Lake Atter, so the low tide seen is nothing unnatural. The air pressure distribution and rapid change in air pressure were such that the otherwise normal phenomenon was exaggerated and took on somewhat larger dimensions.

The 1930 phenomenon made global headlines. The American "The Lewiston Daily Sun" covered the events on 25 July 1930:

BELIEVE IT OR NOT. RIVER FLOWS UP HILL. Linz, Austria, July 25 – (AP) – The amazing spectacle of a river flowing up hill was beheld by vacationists here yesterday.

The waters of the Lake Atter receded and the strange phenomenon caused the River Ager, which has its source in the lake, to start flowing backward.

Many boats on the lake and at the mouth of the river were stranded. The hydrographic institute said the occurrence had no connection with the Italian earthquake.

The same AP story can also be found in the "Hartford Courant" of July 26, 1930 (River Flows Up Hill As Visitors Gape) and even in the "New York Times" on 26 July, 1930, p. 30 (Austrian River in Reverse Leaves Many Boats Stranded).

Witnessed from a height, a seiche should show a more or less straight wave crest racing along the surface of the lake – exactly what was observed from Weyregg. The legend that a large fish wallows in the lake which "foams up wildly and throws up big waves, without even a whispering wind being noticeable" clearly is a tale based on the seiche phenomenon.

Concluding remarks

Lakes with monster sightings tend to attract other motifs as well – in the case of Lake Atter, rumours about hidden Nazi treasure (not uncommon for the German and Austrian lakes of the Alps). The lake certainly has sunken treasure such as Neolithic lake dwellings ("Tages-Post", Linz, 20 May 1910) but the treasure sought in the 1950s was different stuff.

After WWII there were many rumours about sunken Nazi treasure in the lakes of the Alps, and I have collected stories from Lake Chiem and Lake Alat in Germany, Lake Topliz in Austria, and Lake Como in Italy.

We can include Lake Atter in this list. On 24 January 1953, on page 7, the New Zealand "Press" told its readers:

(Rec. 7.30 p.m.) VIENNA, January 23. The wreck of a Nazi Junkers aircraft, said to have been carrying a fortune in gold and platinum when shot down by the Allies, has been found at a depth of 144 feet in Atter Lake, near Salzburg. Divers say it is impossible to lift the plane, but they expect to salvage the treasure. Police are guarding the lake.

The search was still on two years later. As the "Press" reported on 2 September 1955, page 11:

SEARCH FOR NAZI HOARD

View to Weyregg
https://en.wikipedia.org/wiki/Attersee_(lake)#/media/
File:Weyregg_am_Attersee,_vanaf_Attersee_am_Attersee_2e_poging_foto6_2017-08-
11_17.00.jpg
(photo: Michielverbeek, wikimedia.org)

(Rec. 11 p.m.) LINZ (Upper Austria), Sept. 1. Divers will go down to the bed of Lake Atter, near Linz, today in an attempt to find a hoard of Nazi gold reported to have been aboard a German Luftwaffe plane shot down by American fighters towards the end of the last war.

Siegfried Naumann, a 30-year-old diver, has found the plane 240 feet below the lake's surface. The aircraft, according to a local legend and to newspaper reports, held a load of platinum, gold and documents aboard which were being flown, on Hitler's personal orders, to an "alpine redoubt" in which some of. the Nazis hoped to make a last-ditch stand in 1945.

The treasure was not found and the colossal fish never spotted again– or, if it was spotted, the witnesses did not speak with the press. Today, Lake Atter is still one of Austria's beauty spots and attracts millions of visitors each year, but no-one sees a monster. It is forgotten now, no reference to it appears on any of the websites aimed at tourists. However, what happened there in 1902 is an almost replica of what we know from Loch Ness in 1933, and it is a shame the

Kelpie tales

Jonathan Downes

If you are driving to Woolsery from most of the rest of the country, you will have driven along the A39 from Barnstaple, and turned off at the village of Bucks Cross. After about a mile you will pass through a small hamlet of half a dozen houses and then go down a steep hill. At the bottom of the hill, you will drive over a bridge that crosses a small stream, and pass a lone house on the left hand side before driving up a steep hill which leads into Woolsery village itself.

The lone house has just been rebuilt on the site of another house that used to be a mushroom farm (which is totally irrelevant), the hamlet is called Cranford, and the stream is called Cranford Water, and this is quite possibly the longest passage that has ever been written about it.

Now, I don't know how much of this is true, but it was certainly commonly believed amongst the younger people of the village, back when I was one of them.

Whilst lots has been written, and – indeed – filmed as regards the Allied servicemen who became Prisoners of War in Germany and Occupied Europe, far less has been

written and practically nothing filmed about the German servicemen who were taken prisoner by the Allies.

I have read on a number of occasions, however, that the conditions under which prisoners of the Axis Powers were incarcerated by the British were far less draconian than they were for their counterparts in mainland Europe.

It has been reasonably well attested that more than a few German prisoners were granted parole, and were allowed to leave their prison camps in order to work, having given their word of honour that they would not attempt to escape, for many of the landowners in the farming industry. This was particularly useful, because although farmers and agricultural workers were amongst those who were exempt from the call up under the 1939 Military Training Act, the farms were left undermanned, especially as government policy became further reviewed as the war continued and the need for men to join the Armed Forces grew greater.

Women didn't join combat units in those days, and with nearly all the men between the ages of 18-41 years old away at the war, the social structure of the countryside (in particular) was unlike anything that had been seen before. And so, the advent of significant numbers of fit, healthy, blonde and blue-eyed strapping young lads from the rich agricultural lands of the Central European Plain, was a godsend to the farmers who would otherwise never have managed to get the harvest gathered in.

But it wasn't just the farmers who were suffering privations with most of the young men away at the war. It is certain that, despite the political differences between our two countries, romances between these German soldiers and the young (and not so young) women of the region were not unheard of. And so, at least according to what I was told as a boy in my early teens, it was in Woolsery. I don't know where the captured German soldiers, sailors, and airmen, were stationed. And I don't know to which farm they had been assigned. I did know the name of the young woman who took centre stage in this little tragedy, but I was told more or less in confidence, and so I shall not break that confidence even now, half a century later. But, allegedly at least, there had

been a romance between a good-looking young airman and a comely young farmer's daughter. Apparently this romance continued until the end of the war, when, suddenly,

as the rightful young men of the district slowly began to return home, the prisoners of war were shipped in stages back to their fatherland. And this is where the tragedy took place. The young man in question, heartbroken at the thought that he would never see his lover again, hung himself from one of the trees deep in Cranford Wood. His girlfriend is said to have shot herself with her father's shotgun.

She allegedly survived for some years, and, possibly as a result of the injuries sustained, gave birth to an exceedingly unhealthy baby boy, who was one of my patients back when I worked with such people in the early 1980s.

I have no idea if this story is true or not. My friend David and I always set the tragic drama in and around Cranford Water, mainly because, up until they were demolished in late 2020, there were still Nissan Huts in the garden behind the house, and whilst – when we knew it – it was a mushroom farm, David and I imbued it with a desperately sad backstory.

I have no idea whether this story has any relevance whatsoever to the history of weird happenings that have been recorded from the area over the years. As I have written elsewhere, I strongly suspect that the vast majority of "ghosts" are a parapsychological phenomenon, quite possibly in conjunction with a geophysical one, and have very little, if anything, to do with the unquiet spirits of the dead.

When I was a nurse at a local residential hospital for mentally handicapped adults at the beginning of the 1980s, there was a severely handicapped man in his late thirties. His family certainly came from Woolsery, and had the same surname - which I was told in confidence - of that of the young woman who was the major protagonist of this story. His age would have suggested that he was born soon after the cessation of hostilities in 1945, but then again it was a very common North Devon surname.

The story that the young woman concerned survived having shot herself is a romantic one, which conveniently combines the stories of the protagonist of 'Lorna Doone' with the true story of what happened to Unity Mitford; indeed it does so in such a neat manner that I strongly suspect that the events I was told about the tragic love affair between a farmer's daughter and a German POW had been embellished with these two well-known stories – one fictional, the other not. And that is just about it.

Like so many of the events chronicled in this book, and more frustratingly, so many of the events I have spent my professional life trying to untangle, they are lost in a morass of truths, half-truths, lies, parables, and fables; stories with a greater or lesser degree of veracity which were originally told for a variety of reasons to fulfil a variety of different social needs.

Pontius Pilate is said to have asked a rhetorical question, "what is truth?" at Christ's trial. My only answer to this is that truth is highly subjective, and can be a different thing to different people, and that very little in this book illustrates this paradigm as well as the tragic story which may or may not have taken place at Cranford during the second half of World War II.

Back in the mid 1970's, the village was quite isolated and was – in my opinion at least – a far most cohesive society than it has been in more recent years. One of the oft discussed facets of village life is the fact that everybody seems to known everybody else's business, so, when in the early 1970's the rumour went around the young folk that three of their number; the son of a wealthy local businessman, his sister, and his girlfriend had started practising witchcraft, the story spread like wildfire.

Now, you have to remember that as L.P Hartley wrote in 1953, the past is indeed a different country where they do things completely differently. And, in a world where neo-paganism is the fastest growing religion in the country, there are 'Mind, Body and Sprit' sections in every book shop and even some supermarkets, and pretty well everybody has got access to high speed broadband, general knowledge of witchcraft is far more widespread then it was four decades ago.

Back then the only thing that most people knew about the subject was from the novels of Dennis Wheatley, and the occasional pictorial in the 'News of the Screws' featuring Maxine Sanders capering around with her kit off!

As a 14 year old, and further more one coming from a relatively strict Christian household, I knew even less than most people. However, my friend David and I had an advantage over most of our contemporaries; we knew the woodlands around our village intimately. And so, when it was suggested two or three of us go out and investigate the stories suggest that dark Eleusinian rights were taking place in the woods by Cranford Water, we had a pretty good idea of how to proceed.

Truly, our motivation was of the highest order. My mother was a devotee – now discredited book by Dr Margret Murray, especially The Witch Cult in Western Europe which claimed that before Christianity had swept the continent the ancient religion of Western Europe venerated an ancient fertility goddess analogous to Diana or Demeter.

When I had been a child at primary school in Hong Kong, my favourite subject had been Greek Mythology, so the idea that such rights were going on upon my figurative doorstep, was an exciting one, I imagined that the celebrants of these arcane rituals would be wearing long white robes, and possibly playing lyres. I think that a little bit of Ray Harryhausen's imagery from 'Jason and the Argonauts', had probably crept in there somehow as well.

So, on the appointed afternoon, during one of the blisteringly hot late July days which don't seem to happen anymore, four of us, furtively creeping along Red Indian-style walked along the road that still exists from Kennerland Cross toward the Parkham Ash crossroads at the top of Cranford Hill. About three quarters of the way along, there is a dip, and at the bottom of the dip, the stream which eventually runs through Cranford Water passes under the road. Becoming, if anything, even more furtive than before, we followed the course of the stream southwards.

Sneaking along the course of the tiny stream for about twenty minutes we found ourselves approaching a natural clearing where, the previous summer we had all played at being Red Indian Braves. We could hear the sound of voices, and as we crept slowly nearer we could hear that they were chanting. None of us had ever seen 'The Wicker Man' but we knew enough to know that a horrible fiery fate was likely to await any good churchgoing lads who stumbled into the middle of an arcane Pagan ceremony, so we made damn sure that nobody saw us.

This is probably a good thing. Because, as we got closer we could see, instead of the dignified Hellenic ceremony that we had envisaged, there were three teenagers, no more than three years older than we, standing facing each other waving sticks in the air. But the real shock was that they didn't seem to have any clothes on.

All attempts at being serious ethnographic explorers went out the window at the sight of bare flesh and we ran away giggling.

This may mean nothing, or it may mean everything, but it – for me at least – only serves to underline my assertion that Cranford Water, and the surrounding woods are a very strange place indeed.

One night, a salesman acquaintance of mine back in about 1980 was driving home after a long day's toil delivering pet supplies around the pet shops of North Devon. It was a chilly spring night, at about nine o'clock, and I met him by chance in the pub about half an hour later. He had been driving back towards the village, and when he drove over the bridge at Cranford Water a man-sized, in fact an oversized ("It was as big as you, you fat bugger" he said) figure stepped into the road in front of the car.

"There wasn't a hope in hell of stopping in time," he told me in the pub after quaffing several pints of lager and blackcurrant (well it was the beginning of the decade that taste forgot). Therefore, it was probably just as well that the speeding car (he was well known in the village for having a tendency to drive like a maniac) went straight through the figure (and not in an-internal-organs-splattered-all-over-the-Queen's-Highway kind of way).

Years later, in the late 1990s after I had begun to make somewhat of a name for myself as a chronicler of strangeness in all its myriad forms, my father telephoned me with another weird story. Apparently one of his neighbours had also been driving home one evening when they had seen a "golden globe of light" hovering over the road. She put her foot to the floor and juddered to a halt in a far more lady-like manner than the previous witness. At my parents' Golden Wedding celebrations a few months later I met her, and she told me that she could see it hovering in the middle of the road, and it appeared to have occasional spikes of what she described as "golden lightning" sparking out of it. The globe was quite small – about the size of a watermelon – and after a few minutes moved off following the course of the water upstream until it soon vanished from sight. I didn't have to ask her whereabouts when she had seen it, but she confirmed my suspicions. There was, indeed, yet another incident of high strangeness at Cranford Water.

A less benign incident happened in 1975, when one of the girls who lived in the village was walking home from seeing her boyfriend. In those days, there was a bus from Bideford, but it only stopped at Bucks Cross, so anyone wanting to get home after a date had to walk a mile and a half back from the bus stop. In these degenerate days, it is hard to imagine parents allowing their teenage girls to walk such a distance in the middle of the night, but it was commonplace in those days. However, according to the rumours on the school bus, (which were never anything more than rumours) the girl was crossing the bridge at Cranford Water in almost pitch darkness, illuminated only by pale moonlight, when she felt someone come up behind her and wrap their arms around her chest so tight that she could hardly breathe. She struggled free, and swung round to face her assailant.

Needless to say, there was nobody there, and she ran home.

But, the strangest story about Cranford Water is that of the ghost cow. There were two farmers (one who lived up the hill at Cranford, and one who lived in the village itself) who had herds of cattle in fields quite a way from their farms, and both of them (one walking east, and one walking west) had to take their herds across the bridge at Cranford Water in order to take them in for milking.

Both farmers, (long dead) told me independently that something monumentally peculiar was wont to happen during the journey. They would, of course, count their cows before they left the home field, and again when they got back to the farm. Invariably the numbers would tally.

However, if they were to count their herd whilst they were approaching or crossing Cranford Water, there would always be one extra. One of the farmers told me that after several years of this happening he endeavoured to solve the mystery by marking each of them with water-soluble yellow paint, plonking his herd in the middle of the bridge, and separating out all the cows that had been marked. As he was about three quarters through the task, he told me that he felt a sense of panic, became unaccountably dizzy for a few seconds, and lost count of what he was doing. He then heard a cackle of wild laughter and out of the corner of his eye saw a tall human figure running off through the woods.

My friend David and I walked with both farmers on a number of occasions, and tried counting the herd each time. Neither farmer would allow any jiggery-pokery with yellow paint, but – with no word of a lie – each time we counted the herd, there was one extra cow as they crossed the old stone bridge.

But then again, both of us failed our Maths O-Level (that's GCSE for you under forties) with dismal ignominity. And I still count on my fingers to this day.

Now, I give you all these stories for what they are worth, which as none of them are corroboratable (pers. comm. from a bunch of long dead villagers, and the author's father, does not really cut it in academic circles). But it is, I think, of mild interest that so many quasi-fortean occurrences were reported from the same place (and by the way I have ignored a UFO sighting and a big cat sighting from twenty years apart), even though they are apparently disparate and unconnected with each other.

Or are they?

In Scottish folklore there is an entity called the kelpie or waterhorse, which is supposed to be generally malevolent in nature, and which can take on a variety of human and ungulate forms – usually equine. However, if one believes in such things, and I really have no opinion on the matter, then possibly a bovine analogue of the kelpie from the south west of England might not be beyond the bounds of possibility.

But who in their right mind could believe in such a thing? What do you mean, right mind?

But I shall leave you with two stories, one of which can be corroborated without embarrassment.

At Christmas some years ago, Corinna (my long-suffering and remarkably fortean wife who died in 2020) and I were driving home from a Christmas party at the home of some close friends (ironically the sister of `David` whom I mentioned earlier). We were driving back across Cranford Water at about midnight when both of us saw a tall, wild-eyed figure standing by the side of the road gesticulating wildly towards the heavens. Was it some drunk from the village pub wandering home far too late on a midwinter's night? Or could it have been the same phantasm who has scared various hapless passers by and farmers over the years? We both decided that discretion was the better part of valour and didn't stop to find out.

If you don't believe me, you should have asked my wife. She was much nicer than me.

The last story is even less easy to verify. In September 2009 I was in Bideford, preparing for what turned out to be a momentous trip to Ireland. I went into one of the High Street chemists to buy some travel sickness pills. I don't get seasick, but I thought that it was a possibility that one of my companions would do. There, behind the counter was a face that I recognised. I hadn't seen her for thirty five years or more since I caught a glimpse of her, with her brother and another girl through the coppiced trees at Cranford Water.

"Hello. I didn't recognise you with your clothes on", I didn't say. But believe me, I was tempted.

THE GREAT FISH-LIZARDS.

Taken From

EXTINCT MONSTERS.

*A POPULAR ACCOUNT OF SOME OF THE LARGER
FORMS OF ANCIENT ANIMAL LIFE.*

BY

REV. H. N. HUTCHINSON, B.A., F.G.S.,

AUTHOR OF "THE AUTOBIOGRAPHY OF THE EARTH,"
AND "THE STORY OF THE HILLS."

WITH ILLUSTRATIONS BY J. SMIT AND OTHERS.

FIFTH AND CHEAPER EDITION.

LONDON: CHAPMAN & HALL, LD.

1897.

"Berossus, the Chaldæan saith: A time was when the universe was darkness and water, wherein certain animals of frightful and compound forms were generated. There were serpents and other creatures with the mixed shapes of one another, of which pictures are kept in the temple of Belus at Babylon."

—*The Archaic Genesis.*

Visitors to Sydenham, who have wandered about the spacious gardens so skilfully laid out by the late Sir Joseph Paxton, will be familiar with the great models of extinct animals on the "geological island." These were designed and executed by that clever artist, Mr. Waterhouse Hawkins, who made praiseworthy efforts to picture to our eyes some of the world's lost creations, as restored by the genius of Sir Richard Owen and other famous naturalists. His drawings of extinct animals may yet be seen hanging on the walls of some of our provincial museums; and doubtless others still linger among the natural history collections of schools and colleges.

Lazily basking in the sun, when it condescends to shine, and resting his clumsy carcase on the ground that forms the shore near the said geological island at Sydenham, may be seen the old fish-lizard, or Ichthyosaurus, that forms the subject of the present chapter. He looks awkward on land, as if longing to get into his native element once more, and cleave its waters with his powerful tail-fin. His "flippers" seem too weak to enable him to crawl on land. Moreover, the most recent discoveries of Dr. Fraas lead us to conclude that the Ichthyosaur never ventured to leave the "briny ocean" to bask upon the land.

This great uncouth beast presents some curious anomalies in his constitution, being planned on different lines to anything now living, and presenting, as so many other extinct animals do, a mixture, or fusion, of types that greatly puzzled the learned men of the time when his remains were first brought to light, after their long entombment in the Lias rocks forming the cliffs on the coast of Dorset. Some have christened him a "sea-dragon," and such indeed he may be considered. But the name Ichthyosaurus, given above, has received the sanction of high authority, and, moreover, serves to remind us of the fact that, although in many respects a lizard, he yet retains in his bony framework the traces of a remote fishy ancestry. So we will call him a fish-lizard

We remember in our young days the amiable endeavours of Mr. "Peter Parley" to introduce us to the wonders of creation; and his account of the Ichthyosaurus particularly impressed itself on our youthful imagination. How surprised that inestimable instructor of youth would be could he now see the still more wonderful remains that have been brought to light from Europe, Asia, Africa, and America!

The curious quotation given at the head of the present chapter refers to a widespread belief, prevalent among the highly civilised nations of antiquity, that the world was once inhabited by dragons, or other monsters "of mixed shape" and characters. To the student of ancient history traces of this curious belief will be familiar. Sir Charles Lyell refers to such a belief when he says, in his *Principles of Geology*, "The Egyptians, it is true, had taught, and the Stoics had repeated, that the earth had once given birth to some monstrous animals that existed no longer." It may be surprising to some, but it is undoubtedly the fact, that modern scientific truths were partly anticipated by the civilised nations of long ago. Take the ideas of the ancients as interpreted from the records of Egypt, Chaldæa, India, and China; and you will find that our discoveries in geology, astronomy, and ethnology go far to prove that the traditions of these ancient peoples, however derived, after making due allowance for Oriental allegory and poetic hyperbole, are not far from the truth. To the Babylonian tradition of the monstrous forms of life at first created we have already alluded; but in other fields of discovery we find the same foreshadowing of discoveries made in our own day. Take the vast cycles of Egyptian tradition, wherein the stars returned to their places after a circle of constant change, only to start again on their unwearied round; the atomic theory of Lucretius, now expanded and incorporated into modern chemistry; or the philosopher's pregnant saying—*Omne vivum ex ovo* ("Every living thing comes from an egg"). These and other examples might be cited to show how true the old saying is, "There is nothing new under the sun." In the writings of ancient authors may be found singular notices of bones and skeletons found in "the bowels of the earth," which are referred to an imaginary era of long ago, when giants of huge dimensions walked this earth. One is inclined sometimes to wonder whether the old fables of griffins and horrid dragons may not be to some extent based upon the occasional discovery, in former times, of fossil bones, such as evidently belonged to animals the like of which are not to be seen nowadays. (See chaps. xiii. and xiv.)

The illustrious Cuvier, in his day, considered the fish-lizard to be one of the most heteroclite and monstrous animals ever discovered. He said of this creature that it possessed the snout of a dolphin, the teeth of a crocodile, the head and breast-bone of a lizard, the paddles of a whale or dolphin, and the vertebræ of a fish! No wonder that naturalists and palæontologists, whose realm is the natural history of the past, were obliged to make a new division, or order, of reptiles to accommodate the fish-lizard. It is obvious that a creature with such very "mixed" relationships would be out of place in any of the four orders into which living reptiles, as represented by turtles, snakes, lizards, and crocodiles are divided. Here is what Professor Blackie says of the Ichthyosaurus—

> "Behold, a strange monster our wonder engages!
> If dolphin or lizard your wit may defy.
> Some thirty feet long, on the shore of Lyme-Regis,
> With a saw for a jaw, and a big staring eye.
> A fish or a lizard? An ichthyosaurus,
> With a big goggle eye, and a very small brain,
> And paddles like mill-wheels in chattering chorus,
> Smiting tremendous the dread-sounding main."

A glance at our restoration, Plate II., will show that the fish-lizard was a powerful monster, well endowed with the means of propelling itself rapidly through the water as it sought its living prey, to seize it within those cruel jaws. The long and powerful tail was its chief organ of propulsion; but the paddles would also be useful for this purpose, as well as for guiding its course. The pointed head and generally tapering body suggests a capability of rapid movement through the water; and since we know for certain that it fed on fishes, this conclusion is confirmed, for fishes are not easily caught now, and most probably were not easily caught ages ago.

The personal history of the fish-lizard, merely as a fossil or "remain," is interesting; so much so, that we may perhaps be allowed to relate the circumstances of his *début* before the scientific world, in the days of the ever-illustrious Cuvier, to whom we have already alluded. But England had its share of illustrious men, too, though lesser lights compared to the founder of comparative anatomy,—such as Sir Richard Owen, on whom the mantle of his friend Cuvier has fallen; Conybeare, De la Beche, and Dean Buckland.

PLATE II.

FISH-LIZARDS.

Ichthyosaurus communis.	*Ichthyosaurus tenuirostris.*
Length about 22 feet.	Fishes, *Dapedius*, etc. A smaller species.

These scientific men, aided by the untiring labours of many enthusiastic collectors of organic remains, have been the means of solving the riddle of the fish-lizard, and of introducing him to the public. By this time there is, perhaps, no creature among the host of Antediluvian types better known than this reptile.

The remains of fish-lizards have attracted the attention of collectors and describers of fossils for nearly two centuries past. The vertebræ, or "cup-bones," as they are often called, of which the spinal column was composed, were figured by Scheüchzer, in an old work entitled *Querelæ Piscium*; and, at that time, they were supposed to be the vertebræ of fishes. In the year 1814 Sir Everard Home described the fossil remains of this creature, in a paper read before the Royal Society, and published in their *Philosophical Transactions*. This fossil was first discovered in the Lias strata of the Dorsetshire coast. Other papers followed till the year 1820. We are chiefly indebted to De la Beche and Conybeare for pointing out and illustrating the nature of the fish-lizard; and that at a time when the materials for so doing were far more scanty than they are now. Mr. Charles König, Mr. Thomas Hawkins, Dean Buckland, Sir Philip Egerton, and Professor Owen have all helped to throw light on the structure and habits of these old tyrants of the seas of that age, which is known as the Jurassic period. They lived on, however, to the succeeding or Cretaceous period, during which our English chalk was forming; but the Liassic age was the one in which they flourished most abundantly, and developed the greatest variety.

In the year 1814 a few bones were found on the Dorsetshire coast between Charmouth and Lyme-Regis, and added to the collection of Bullock. They came from the Lias cliffs, undermined by the encroaching sea. Sir Everard's attention being attracted to them, he published the notices already referred to. The analogy of some of the bones to those of a crocodile, induced Mr. König, of the British Museum, to believe the animal to have been a saurian, or lizard; but the vertebræ, and also the position of certain openings in the skull, indicated some remote affinity with fishes, but this must not be pressed too far. The choice of a name, therefore, involved much difficulty; and at length he decided to call it the *Ichthyosaurus*, or fish-lizard. Mr. Johnson, of Bristol, who had collected for many years in that neighbourhood, found out some valuable particulars about these remains. The conclusions of Dean Buckland, then Professor of Geology at Oxford, led Sir Everard to abandon many of his former conclusions. The labours of the learned men of the day were greatly assisted by the exertions of Miss Anning, an enthusiastic collector of fossils. This lady, devoting herself to science, explored the frowning and precipitous cliffs in the neighbourhood of Lyme-Regis, when the furious spring-tide combined with the tempest to overthrow them, and rescued from destruction by the sea, sometimes at the peril of her life, the few specimens which originated all the facts and speculations of those persons whose names will ever be remembered with gratitude by geologists.

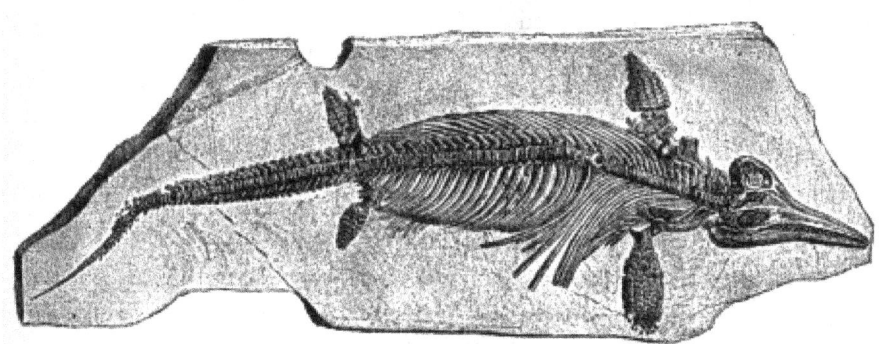

FIG. 3.—*Ichthyosaurus intermedius.*

Probably our readers are already more or less familiar with the drawings of the fossilised remains of Ichthyosauri to be seen in almost every text-book of geology. (Fig. 3 is from Owen's *British Fossil Reptiles*.) But we recommend all who take an interest in the world's lost creations to pay a visit to the great Natural History Museum, at South Kensington. The fossil reptile gallery contains a magnificent series of Ichthyosauri, about thirty in number. Of these a large number were obtained through the exertions of the late Mr. T. Hawkins, a Somersetshire gentleman, who was a most ardent collector of fossil reptiles, and who devoted himself with great enthusiasm and unsparing energy to the acquisition of a truly splendid collection of these most interesting relics of the past. Nearly sixty years ago he arranged for the purchase of his treasures by the authorities of the British Museum, and thus his collection became the property of the nation.

His specimens were figured and described by him in two large folio volumes. The first was published in 1834, under the title, *Memoirs of the Ichthyosauri and Plesiosauri*; his second, with the same plates, in 1842, under the quaint title of *The Book of the Great Sea-Dragons*. The large lithographic drawings of his fine specimens were beautifully executed by Scharf and O'Neil. The plates are the only really valuable part of these two curious and ill-written books. His descriptions are not of much value, and his pages are encumbered with a vast amount of extraneous matter. The author is immensely proud of his collection, and his vanity is conspicuous throughout. Instead of confining himself to descriptions of what he found, and how he found them, he continually wanders into all sorts of subjects that are, to say the least, irrelevant. In one place he introduces ancient history and mythology; in another, Old Testament chronology; in another, the unbelieving spirit of the age; and here and there indulges in vague unphilosophical speculations. Altogether his two volumes are a curious mixture of bigotry, conceit, and unrestrained fancy, and they afforded to the present writer no small amusement. One rises from the perusal of such men's writings with a strong sense of the contrast between the humble and patient spirit in which our great men of to-day, such as Professor Owen, study nature and record their observations, and the vague, conceited outpourings of some old-fashioned writers.

Mr. Hawkins tells us that his youthful attention was directed to the Lias quarries, near Edgarly, in Somersetshire, in consequence of some strange reports. It was said that the bones of giants and infants had, at distant intervals, been found in them. These quarries he visited, and, by offers of generous payment, induced the workmen to keep for him all the remains they might find. In this way he finally obtained the co-operation of all the quarrymen in the county.

Mr. Hawkins thus expresses his delight on obtaining an Ichthyosaurus which was pointed out to him by Miss Anning, near the church at Lyme-Regis, in the year 1832: "Who can describe my transport at the sight of the colossus? My eyes the first which beheld it! Who shall ever see them lit up with the same unmitigated enthusiasm again? And I verily believe that the uncultivated bosoms of the working men were seized with the same contagious feeling; for they and the surrounding spectators waved their hats to an 'Hurra!' that made hill and mossy dell echoing ring."

This specimen, however, got sadly broken in its fall from the cliff; but in time he put all the pieces together again. Speaking of his own collection, he says, "This stupendous treasure was gathered by me from every part of England; arranged, and its multifarious features elaborated from the hard limestone by my own hands. A tyro in collecting at the age of twelve years, I then boasted of all the antiquities that were come-at-able in my neighbourhood, but, finding that everybody beat my cabinet of coins, I addressed myself to worm-eaten books, and last to fossils." Before he was twenty years of age he had obtained a very fine collection of organic remains.

When, however, he complains of the Philistine dulness and stupidity of quarrymen, who often, in their ignorance, break up finds of almost priceless value, we can fully sympathize.

In general contour the body of the fish-lizard was long and tapering, like that of a whale (see <u>Plate II.</u>). It probably showed no distinct neck. The long tail was its chief organ of propulsion. We notice two pairs of fins, or paddles; one on the fore part of the body, the other on the hinder part, like the pectoral and abdominal fins of a fish. The skin was scaleless and smooth, or slightly wrinkled, like that of a whale. No traces of scales have ever been found; and if such had existed, they would certainly have been preserved, since those of fishes and crocodiles of the Jurassic period have been found in considerable number and variety. It is therefore safe to conclude that such were absent in this case. In the Lias strata, at least, the specimens are often preserved with most wonderful completeness.

The long and pointed jaws are a striking feature of these animals. The eyes were very large and powerful, and specially adapted, as we shall see presently, to the conditions of their life.

t might, perhaps, be asked whether the fish-lizards breathed, like fishes, by means of gills. That question can easily be answered; for if they had possessed gills for taking in water and breathing the air dissolved therein, they would reveal the fact by showing a bony framework for the support of gills, such as are to be found in all fishes. These structures, known as "branchial arches," are absent; therefore the fish-lizards possessed lungs, and breathed air like reptiles of the present day. Their skulls show where the nostrils were situated; namely, near the eyes, and not at the end of the upper jaw-bone. There are also passages in the skull leading from the nostrils to the palate, along which currents of air passed on their way to the lungs. Being air-breathers, they would be compelled occasionally to seek the surface of the sea, in order to obtain a fresh supply of the life-giving element—oxygen; but, being cold-blooded and with a small brain, needing a much less supply of oxygen for its work, the fish-lizards had, like fishes, this advantage over whales, which are warm-blooded—that their stern-propeller, or tail-fin, could take the form best adapted for a swift, straight-forward course through the water.

In the whale tribe the tail-fin is horizontal; and this is so on account of their need, as large-brained, warm-blooded air-breathers, of speedy access to the atmospheric air. Were it otherwise, they would not have the means of rising with sufficient rapidity to the surface of the sea; for they have only one pair of fins. But the fish-lizards had two pairs of these appendages, and the hinder or pelvic pair no doubt were of great service in helping the creatures to come up to the surface when necessary.

Thus we see that the whale, with its one pair of paddles, has a tail specially planned with a view to rapid vertical movement through the water; while in the fish-lizards, who did not require to breathe so frequently, the tail-fin was planned with a view to swift and straight movement forward as they pursued their prey, and they were compensated by having bestowed upon them an extra pair of paddles. Thus we learn how one part of an animal is related to and dependent upon another, and how they all work together with the greatest harmony for certain definite purposes

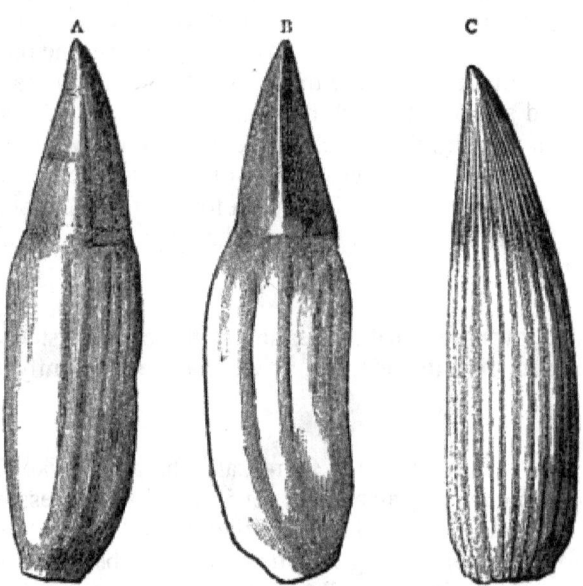

FIG. 4.—(A) Lateral and (B) profile views of a tooth of *Ichthyosaurus platyodon* (Conybeare), Lower Lias, Lyme Regis, Dorsetshire, (C) Tooth of *Ichthyosaurus communis* (Conybeare), Lower Lias, Lyme Regis, Dorset.

These great marine predaceous reptiles literally swarmed in the seas of the Lias period, and no doubt devoured immense shoals of the fishes of those times, whose numbers were thus to some extent kept down. There is clear proof of this in the fossilised droppings—known as "coprolites,"—which show on examination the broken and comminuted remains of the little bony plates of ganoid fishes that we know were contemporaries of these reptiles. Probably young ones were sometimes devoured too.

It was in the period of the Lias that fish-lizards attained to their greatest development, both in numbers and variety; and the strata of that period have preserved some interesting variations. It will be sufficient here to point out two, namely, Ichthyosaurus tenuirostris—an elegant little form, in which the jaws, instead of being massive and strong, were long and slender like a bird's beak; and also Ichthyosaurus latifrons (Fig. 5), with jaws still more bird-like. Our artist has attempted to show the former variety in our illustration (Plate II.). A most perfect example of this pretty little Ichthyosaur, from the Lower Lias of Street in Somerset, has recently been presented to the National Collection at South Kensington by Mr. Alfred Gillett, of Street, and may be seen there. In this group of fish-lizards the eyes are relatively larger, and we should imagine that they were very quick in detecting and catching their prey; their paddles also have larger bones.

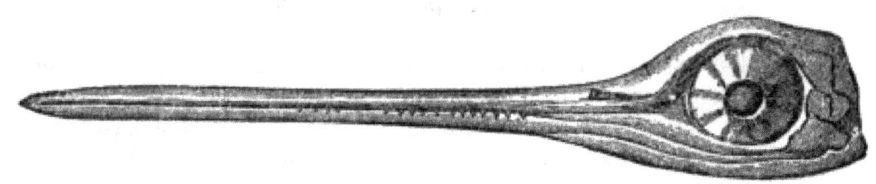

FIG. 5.—Skull of *Ichthyosaurus latifrons*.

There is a remarkably fine specimen at Burlington House, in the rooms of the Geological Society, of an Ichthyosaurus' head, which the writer found, on measuring, to be about five feet six inches long. A cast of this head is exhibited at South Kensington. The largest of the specimens in the National Collection is twenty-two feet long and eight feet across the expanded paddles; but it is known that many attained much greater dimensions. Judging from detached heads and parts of skeletons, it is probable that some of them were between thirty and forty feet long. A specimen of Ichthyosaurus platyodon in the collection of the late Mr. Johnson, of Bristol, has an eye-cavity with a diameter of fourteen inches. This collection is now dispersed.

With regard to their habits, Sir Richard Owen concludes that they occasionally sought the shores, crawled on the strand, and basked in the sunshine. His reason for this conjecture (which, however, is not confirmed by Dr. Fraas's recent discoveries) is to be found in the bony structure connected with the fore-paddles, which is not to be found in any porpoise, dolphin, grampus, or whale, and for want of which these creatures are so helpless when left high and dry on the shore. The structure in question is a strong bony arch, inverted and spanning across beneath the chest from one shoulder to the other. A fish-lizard, when so visiting the shore for sleep, or in the breeding season, would lie or crawl, prostrate, with its under side resting or dragging on the ground—somewhat after the manner of a turtle.

It is, perhaps, hardly necessary to remark that whales are not fishes, but mammals which have undergone great change in order to adapt themselves to a marine life. Their hind limbs have practically vanished, only a rudiment of them being left.

t is a curious fact that this bony arch resembles the same part in those singular and problematical mammals, the Echidna and the Platypus, or duck-mole.

The enormous magnitude and peculiar construction of the eye are highly interesting features. The expanded pupil must have allowed of the admittance of a large quantity of light, so that the creature possessed great powers of vision.

The organic remains associated with fish-lizards tell us that they inhabited waters of moderate depth, such as prevails near a coast-line or among coral islands. Moreover, an air-breathing creature would obviously be unable to live in "the depths of the sea;" for it would take a long time to get to the surface for a fresh supply of air

Perhaps no part of the skeleton is more interesting than the curious circular series of bony plates surrounding the iris and pupil of the eye. The eyes of many fishes are defended by a bony covering consisting of two pieces; but a circle of bony overlapping plates is now only found in the eyes of turtles, tortoises, lizards, and birds, and some alligators. This elaborate apparatus must have been of some special use; the question is—What service or services did it perform? Here, again, we find answers suggested by Owen and Buckland. It would aid, they say, in protecting the eye-ball from the waves of the sea when the creature rose to the surface, as well as from the pressure of the water when it dived down to the bottom—for even at a slight depth pressure increases, as divers know. But it appears that the ring of bony plates fulfilled a yet more important office, thereby enabling the fish-lizards to play admirably their part in the world in which they lived, and to succeed in the struggle of life; for even in those remote days there must have been, as now, a keen competition among all animals, so that the victory was to those that were best equipped.

Would it not be an advantage for them to have the power of seeing their finny prey whether near or far? Certainly it would; and so we are told that, by bringing the plates a little nearer together, and causing them to press gently on the eye-ball, so as to make the eye more convex—that is, bulging out—a nearer object would be the better discerned. On the other hand, by relaxing this pressure, thus enlarging the aperture of the pupil and diminishing the convexity, a distant object would be focussed upon the retina. In this manner some birds alter the focus of their eyes while swooping down on their prey.

What a wonderful arrangement! We often hear of people having two pairs of spectacles—with lenses of different curvature—one for reading, and the other for seeing more distant objects than a book held in the hand. But here is a creature that possessed an apparatus far more simple and effective than that supplied by the optician! Dr. Buckland, speaking of these "sclerotic plates," as they are called, says they show "that the enormous eye of which they formed the front was an optical instrument of varied and prodigious power, enabling the Ichthyosaurus to descry its prey in the obscurity of night and in the depths of the sea." But the last expression must be taken in a limited sense (see <u>Fig. 6</u>).

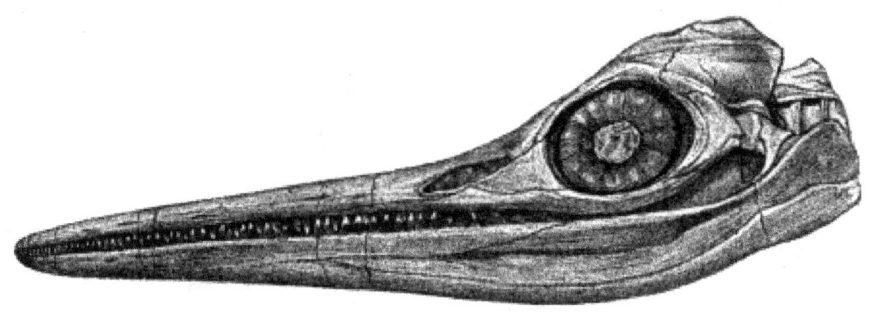

\#

Fig. 6.—Head of *Ichthyosaurus platyodon.*

It might well be supposed that no record had been preserved from which we could learn anything about the nature of the skin of our fish-lizard; but even this wish has been partly fulfilled, to the delight of all geologists. Certain specimens have been obtained, from the Lias of England and Germany, that show faithful impressions of the skin that covered the paddles. A specimen of this nature has lately been presented to the national treasure-house at South Kensington by Mr. Montague Brown. On the inner side of the paddle was a broad fin-like expansion, admirably adapted to obtain the full advantage of the stroke of the limb in swimming.

Mr. Smith Woodward informs the writer that specimens have lately been found near Würtemberg, with evidence of a triangular fin on the back. Plate II. has been redrawn for this edition, to make it more in harmony with Dr. Fraas's discoveries.

Speaking of the limbs, it should be mentioned that the bones of each finger, instead of being elongated and limited in number to three in each of the five fingers, are polygonal in shape and arranged in as many as seven or eight rows, while those of each finger are exceedingly numerous. Thus the whole structure forms a kind of bony pavement which must have been very supple. Such a limb would be one of the most efficient and powerful swimming organs known in the whole animal kingdom. In whales the fingers of the flippers are of the usual number, namely, five. Some species of fish-lizards had as many as over a hundred separate little bones in the fore-paddle.

Another question naturally suggests itself: Were they viviparous, or did they lay eggs like crocodiles? This question seems to have been answered in favour of the first supposition; and in the following interesting manner. It not infrequently happens that entire little skeletons of very small individuals are found under the ribs of large ones. They are invariably uninjured, and of the same species as the one that encloses them, and with the head pointing in one direction. Such specimens are most probably the fossilised remains of little fish-lizards, that were yet unborn when their mothers met with an untimely end In some cases, however, they may be young ones that were swallowed.

The jaws of these hungry formidable monsters were provided with a series of formidable teeth—sometimes over two hundred in number—inserted in a long groove, and not in distinct sockets, as in the case of crocodiles. In some cases, sixty or more have been found on each side of the upper and lower jaws, giving a total of over two hundred and forty teeth! The larger teeth may be two inches or more in length.

The jaws were admirably constructed on a plan that combined lightness, elasticity, and strength. Instead of consisting of one piece only, they show a union of plates of bone, as in recent crocodiles. These plates are strongest and most numerous just where the greatest strength was wanted, and thinner and fewer towards the extremities of the jaw. A crocodile, Sir Samuel Baker says, in his *Wild Beasts and their Ways*, can bite a man in two; and no doubt our fish-lizard would have been glad to perform the same feat! But in his pre-Adamite days the opportunity did not present itself.

The spinal column, or backbone, with its generally concave vertebræ, must have been highly flexible, as is that of a fish, especially the long tail which the creature worked rapidly from side to side as it lashed the waters

The hollows of these concave vertebræ must have been originally filled up with fluid forming an elastic bag, or capsule. To get a clearer idea of this, take a small portion of the backbone of a boiled cod, or other "bony" fish, and you will see on pulling it to pieces, the white, jelly-like substance that fills up the hollows between the vertebræ. In this way Nature provides a soft cushion between the joints, that allows of a certain amount of movement, while, at the same time, the column holds together. The backbone of a fish may not inaptly be compared to a railway train. Each of the carriages represents a vertebra, and the buffers act as cushions when the train is bent in running round a curve. After all, we must learn from Nature; and many of the greatest mechanical and engineering triumphs of to-day are based upon the methods used by Nature in the building up and equipment of vegetable and animal forms of life

It may, perhaps, be inquired whether there is any evidence for the existence of a tail-fin, such as is shown in our illustration. To this it may be replied that the presence of such an appendage is as good as proved by a certain flattening of the vertebræ at the end of the tail, detected by Owen. The direction of this flattening is from side to side, and therefore the tail-fin must have been vertical, like that of a fish. In one specimen Sir Richard Owen has detected as many as 156 vertebræ to the whole body.

Our description of the fish-lizard has, we trust, been sufficient—although not couched in the language used by men of science—to give a fair idea of its structure and habits.

In conclusion, a few words may be said about the ancestry and life-history of these ancient monsters. Palæontologists have good reason to believe that they were descended from some early form of land reptile. If so, they show that whales are not the first land animals that have gone back to the sea, from which so many forms of life have taken their rise

During the long Mesozoic period fish-lizards played the part that whales now play in the economy of the world; and they resembled the latter, not only in general shape, but in the situation of the nostrils (near the eye), and in their teeth and long jaws. But these curious resemblances must not be interpreted to mean that whales and fish-lizards are related to each other. They only show that similar modes of life tend to produce artificial resemblances—just as some whales, in their turn, show a superficial resemblance to fishes.

With regard to the particular form of reptile from which the fish-lizard may have been derived, no certain conclusion has at present been arrived at. This is chiefly from want of fuller knowledge of early forms, such as may have existed in the previous periods known as the Carboniferous and Trias. But there are certain features in the skulls, teeth, and vertebræ that suggest a relationship with the Labyrinthodonts, or primæval salamanders that flourished during the above periods, or at least from amphibians more or less closely allied to them. They cannot by any possibility be regarded as modified fishes; for fishes have gills instead of lungs.

The fish-lizards played their part, and played it admirably; but their days were numbered, and the place they occupied has since been taken by a higher type—the mammal. As reptiles, they were eminently a success; but, then, they were only reptiles, and therefore were at last left behind in the struggle for existence, until finally they died out, at the end of the Cretaceous period, when certain important geographical and other changes took place, helping to cause the extinction of many other strange forms of life, as we shall see later on

They had a wide geographical range; for their remains have been discovered in Arctic regions, in Europe, India, Ceram, North America, the east coast of Africa, Australia, and New Zealand.

In American deposits they are represented by certain toothless forms, to which the name Sauranodon ("toothless lizard") has been given. These have been discovered by Professor Marsh, in the Jurassic strata of the Rocky Mountains. They were eight or nine feet long, and in every other respect resembled Ichthyosaurs. As we have endeavoured to indicate in our illustration, the fish-lizards flourished in seas wherein animal, and doubtless vegetable life was very abundant. Any one who has collected fossils from the Lias of England will have found how full it is of beautiful organic remains, such as corals, mollusca, encrinites, sea-urchins, and other echinoids, fishes, etc.

The climate of this period in Europe was mild and genial, or even semi-tropical. Coral reefs and coral islands varied the landscape. There is just one more point of interest that ought not to be omitted; it refers to the manner in which these reptiles of the Lias age met their deaths, and were thus buried up in their rocky tombs. Sir Charles Lyell and other writers point out that the individuals found in those strata must have met with a sudden death and quick burial; for if their uncovered bodies had been left, even for a few hours, exposed to putrification and the attacks of fishes at the bottom of the sea, we should not now find their remains so completely preserved that often scarcely a single bone has been moved from its right place. What was the exact nature of this operation is at present a matter of doubt.

Mystery giant salamanders of the world part one

Shane Lea and Richard Muirhead

Welcome to a new investigation into another obscure group of cryptids. This time we are looking at mystery giant salamanders. Deep in the heart of Sasquatch country, that is to say, parts of California and Washington state, in

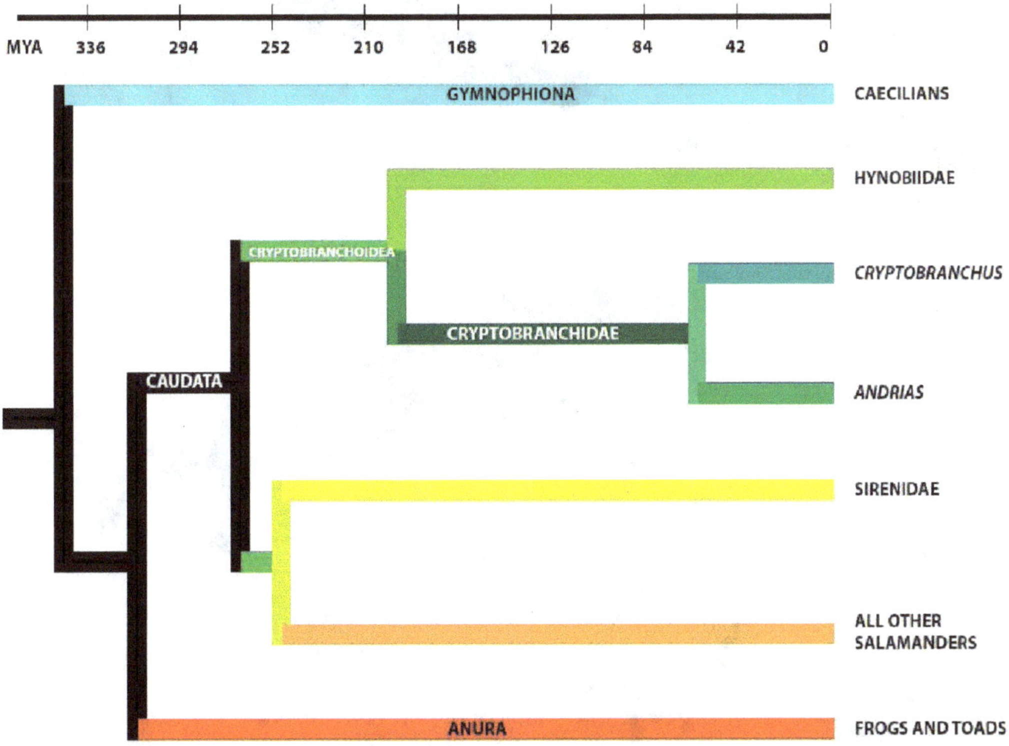

| MYA | 336 | 294 | 252 | 210 | 168 | 126 | 84 | 42 | 0 |

GYMNOPHIONA — CAECILIANS

HYNOBIIDAE

CRYPTOBRANCHOIDEA

CRYPTOBRANCHUS

CRYPTOBRANCHIDAE

ANDRIAS

CAUDATA

SIRENIDAE

ALL OTHER SALAMANDERS

ANURA — FROGS AND TOADS

isolated pools and streams, resides the Trinity Alps giant salamander, if several eyewitnesses are to be believed.

According to "official" zoology the only giant salamanders in the world are in China and Japan (the Cryptobranchidae) and the hellbenders of the eastern and central United States. One of the best summaries of these reports was provided by Loren Coleman in *Fortean Times* #103, October 1997 (`Promises of Giants.`)

"Keystone Mine Attorney Frank Griffith was one of the first on record

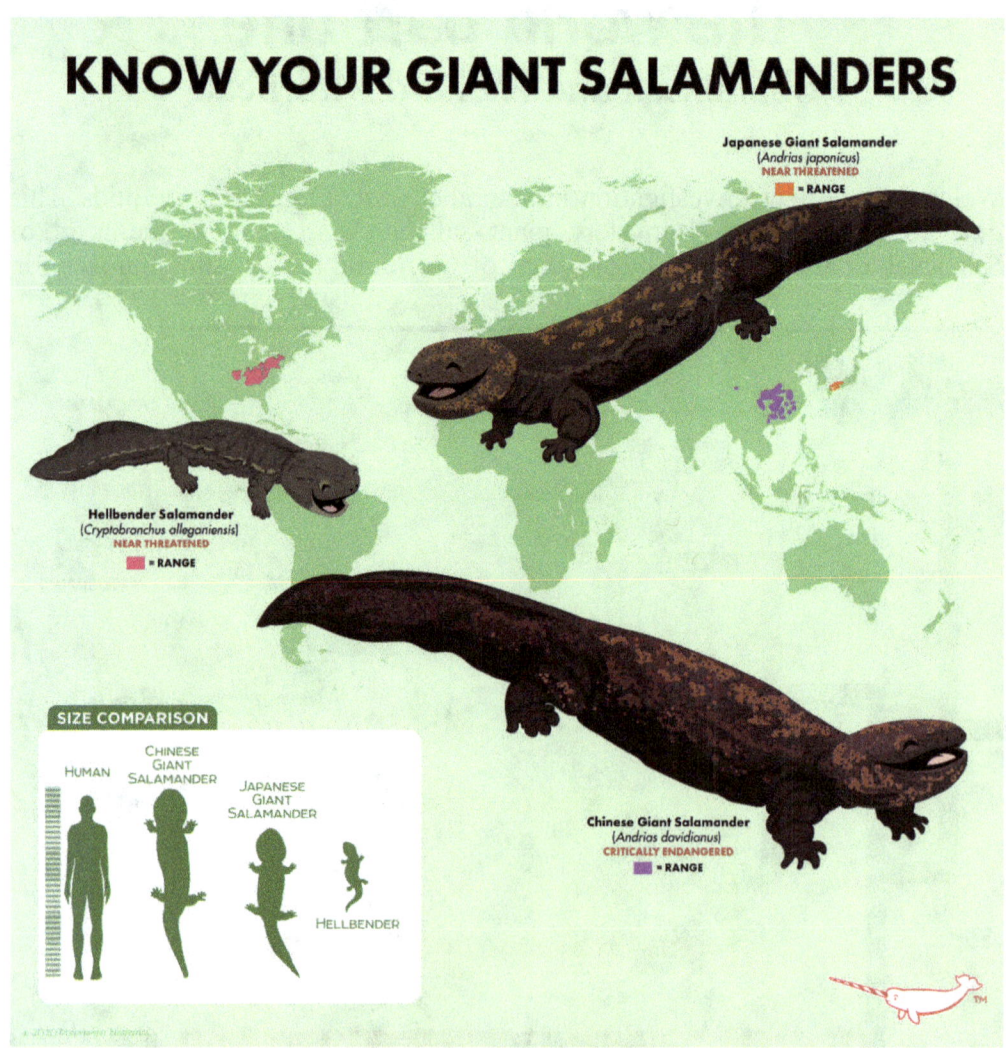

to see giant salamanders in the Trinity Alps. It was during the 1920s, while he was hunting deer near the head of the New River. At the bottom of a lake, he spotted five, some 5-9ft (1.5m-2.75m) long. He caught one on a hook, but he could not pull it out. Hearing Griffith`s story, biologist Thomas Rodgers made four unsuccessful trips in 1948 to try to locate the creatures. He thought they might be Pacific giant salamanders, *Dicamptodon,* (pictured above) even though these are never bigger than a foot long. Perhaps they were a relict population of *Megalobatrachus,* the Asiatic giant salamander of the family Cryptobrachidæ, reputed to measure 5-6ft (1.5-1.8m)…" [1]

In *Copeia*, June 1951, number 2, pp 179-180, George S. Myers (`Asiatic Giant Salamander Caught in the Sacramento River`) wrote:

"In 1939 or 1940, I received a message to the effect that a Sacramento commercial fisherman had caught a "strange" animal which he had alive in his apartment. The animal was a fine

Megalobatrachus (unquestionably identified generically by its closed gill openings), in perfect condition, alive in a wooden trough suspended in the bath tub, where I handled and examined it carefully for about 30 minutes. It was between 25 and 30 inches in length. Its captor said he found it in one of his catfish traps, set in the Sacramento River "below Sacramento but above the delta". Probably this was not far from Walnut Grove... The source of the specimen is, of course, unknown. Its strange coloration even suggested the possibility of a native Californian *Megalobatrachus,* which would not be zoogeographically surprising, but no other captures have been reported..." [2]

Coleman pointed out:

"Nevertheless, it was a very different colour to those of the Japanese and Chinese species. It was dark brown, not slate grey as in the Asian types, and it had dull yellow spots, not ones that were darker grey as is found on the known giant salamanders..." [3]

A few years after this a man called Vern Harden caught one of a dozen giant salamanders he had seen in a lake in the Trinity Alps. He measured it at 8ft 4 inches then released it. In 1960, a Father Hubbard...

"...declared he had definitely established that the huge amphibians were in the Trinity region and would soon leave on another expedition..." [4]

But no record of this expedition exists. Later in 1960 millionaire cryptozoologist Tom Slick took up an interest in the Trinity Alps giant salamanders but was not successful in finding any. According to *Copeia* 3 (September 1962)

REPORT OF GIANT SALAMANDER IN CALIFORNIA
by Thomas Rodgers

The Sacramento River specimen witnessed by Myers was brought from "somewhere in China" by Wong Hong an odd fish fancier. In September, 1960, Nathan Cohen, Thomas Rodgers and Robert Stebbins mounted an expedition for a hunt for the giant salamander and in 1962 Thomas Rodgers mounted an expedition but only collected "about a dozen *Dicamptodon,* the largest of which was $11\frac{1}{2}$in (30cm) long. Rodgers` final comment:

"It is hoped that this evidence will kill rumours about any giant salamanders (much less *Megalobatrachus*) in the Trinity Mountains of California..." [5]

However in 1997 Kyle Mizokami mounted an expedition to look for these cryptids but sadly returned with no further evidence. In 2021 there was another expedition to look for the Trinity Alps giant salamanders involving Ken Gerhard, Jamie Wayne, Jim Whitehead, Jason Kauntze and Daniel Alan Jones. It was not successful. It is possible that the giant salamanders observed in the 1920s were still alive in the 1990s because according to the Giant Salamander Animal Facts - AZ Animals web-site - one giant salamander lived seventy-two years.

According to "Anonymous" in a reply to Raheel Mughal`s post `The Trinity Alps Giant Salamander` on the *Cryptozoology Online* blog of January 12th 2011: "Giant Hellbenders" are reported in the Ohio River valley in Indiana, Ohio and Kentucky." Furthermore in a post on the Facebook Trinity Alps Giant Salamander Information and Discussion group by Daniel Alan Jones on August 4th 2022, citing a book *The Best of Texas Folk and Folklore 1916-1954:*

"Along the New Mexican border is related a story of an enormous and extremely ferocious animal that formerly in- habited one of the salt lakes in Andrews County, Texas. It was an amphibious animal "resembling a great water dog" and was said to feed almost exclusively on ducks and other waterfowl. It was said to make a terrific noise as it plowed its way through the water in the early morning, causing the ducks to arise in great flocks from the surface of the lake. It even put to hurried flight parties of gunners who became frightened at its ferocious appearance. A description of the animal given in a copy of a small plains newspaper of twenty-five years ago (unfortunately since lost) reminded one of the great prehistoric animals. "The "great water-dog' has not been reported in recent years."

Loren Coleman, in a post on the above mentioned Facebook group on March 7th, 2023, mentioned the following:

"The fossil record of North America does indeed include giant salamanders analogous to what's reported in the Cryptozoological literature, I'm not saying I support giant extant salamanders I'm America (as we currently lack the evidence for that). What I am saying is it's plausible"

According to Jim Whitehead on this Facebook group in early January 2023:

"The Kiowa refer to "zemo'hgú-ani" a large predatory horned lizard that lives in rivers. That one is the most well known but some of the other tribes have seen them as well. I have about 10-12 sightings of what locally are called the "walking catfish" which sounds like a large neotenic salamander about 2-3 feet long. And interestingly enough, *Time Life* magazine did a article about monster legends and folklore in the US during the 50's.

They placed a horned lizard like creature over western Oklahoma

and called it the "water monster", so apparently it was more well known in the past."

The Kiowa were originally indigenous to the Great Plains.

According to Karl Shuker`s *Shuker Nature* blog of May 20th, 2015, `A giant mystery salamander from California and a giant mystery salamander from Vietnam`, a woman he called "Prunella" in 2005 saw a salamander or newt, reddish brown and mottled and 4-5ft long in Redwood Park in Arcata, California.

Karl sent her a video of a hellbender and she said it was "so much bigger than the hellbender." In an e-mail to Karl she compared it with the Californian coastal salamander *Dicamptodon tenebrosus*.

EDITOR'S NOTE: I found, to my interest, that there are actually four distinct species of Pacific giant salamanders, which adds weight to the argument that a particularly large species (or morph of a currently known species) pay exist awaiting discovery.

I found two old stories of "water dogs" which may have been giant salamanders.

Image	Scientific name	Common Name	Distribution
	Dicamptodon aterrimus	Idaho giant salamander	forested watersheds from lake Coeur d'Alene to the Salmon River, and in two locations in Montana around Mineral County, Idaho
	Dicamptodon copei	Cope's giant salamander	Olympic Peninsula to northern Oregon
	Dicamptodon ensatus	California giant salamander	Northern California
	Dicamptodon tenebrosus	Coastal giant salamander	Northern California, Oregon, Washington, and southern British Columbia.

The first is from the *Grant County Herald*, Wisconsin, of November 10th, 1920:

> "CATCHES WATER DOG Will Strickland of Galena made an unusual catch on his fish line while down the river Sunday. Instead of a fish he caught a water pup or water dog. It is seemingly a fish with four legs, or a cross between a fish and an alligator…"

The second story is from *The Daily Alaska Empire* of October 17th, 1934:

"IT IS A WATER DOG,NOT LIZARD,CLAIMS ANDERSON... Mr Anderson believes the specimen is an amphibian ,known as a newt or water dog. The only others he has seen in Alaska were along streams Ketchikan…"

EDITOR'S NOTE: There is a group of aquatic animals in North America called water

Image	Scientific name	Common Name	Distribution
	Necturus alabamensis Viosca, 1937	Alabama waterdog	Alabama.
	Necturus beyeri Viosca, 1937 synonym: N. lodingi Viosca, 1937	western waterdog (formerly the Gulf Coast waterdog) or Mobile mudpuppy. These two names have been recognised as independent species in the past.[3][11]	Alabama, Louisiana, Mississippi, and Texas.
	Necturus lewisi Brimley, 1924	Neuse River waterdog	North Carolina.
	Necturus louisianensis Viosca, 1938	Red River waterdog. Sometimes considered a subspecies of N. maculosus.[8][10]	southeastern Kansas, southern Missouri, northeastern Oklahoma, Arkansas, and northcentral Louisiana.
	Necturus maculosus (Rafinesque, 1818)	common mudpuppy	southern section of Canada, as far south as Georgia
	Necturus moleri Guyer et al., 2020	Apalachicola waterdog	southeastern Alabama, the Panhandle of Florida, and southwestern to north-central Georgia.
	Necturus mounti Guyer et al., 2020	Escambia waterdog	southern Alabama and the Panhandle of Florida.
	Necturus punctatus (Gibbes, 1850)	dwarf waterdog	from southeastern Virginia to southcentral Georgia.

dogs. They are in the genus *Necturus*, the most well-known of which is the Mudpuppy. They are aquatic salamanders and there are quite a few different species spread across North America. When I was in Canada in the late 1970s, I became quite familiar with several of these extraordinary animals. However, I am not using this as an excuse to fob Richard and Shane off. There have been stories of particularly shaped and extra large water dogs for many years, and I think that they are quite possibly responsible for some of the North American mystery salamanders.

There is a different category of cryptids in the United States which might be a type of mystery giant salamander or mud puppy. They are large and pink and were investigated by the late American cryptozoologist Mark Hall, in his *Wonders* Volume 1 number 4 (December 1992).

In 'Sobering Sights of Pink Unknowns', Hall states:

> "In my book *Natural Mysteries* I discussed the Giant Pink Lizards of Ohio, which appeared to be the larval stage of a giant salamander still unrecognized by establishment scientists. Two centuries ago they were common in the area of Scippo Creek."

Hall then goes on to describe pink alligators in Florida in 1976, and a case near Charleston, S. Carolina of a pinky-orange creature that might have been a large (c. 6ft long) salamander or hellbender. This was around 1928.

In around 1972, the famous cryptozoology author Ivan Sanderson, saw a pinky creature in a swamp near his home in New Jersey that he compared to a Tatzelwurm, which is a cryptid we will come onto in Part Two. Finally there is an account of a pink unknown from Vermont.

Writing in his newspaper column "Fishy Tales" in the *Rutland Herald* of August 10th, 1980, Charles Spencer mentioned a "pink crocodile" which may actually have been a larval stage giant salamander or mudpuppy in the headwaters of the Clarendon River. In British Columbia, in Canada, there is a large mystery black cryptid which may be a giant salamander known as a "black alligator."

According to one web site—Cryptid Profile: Canadian Alligator (aka Pitt Lake Lizard):

> "On October 10th, 1900, on the shore of Kootenay Lake, George Goudereau watched from a distance as a 12ft creature that roughly resembled an alligator crawled out of Crawford Bay and made its way towards a pile of rotten compost that had been collecting on land. George starred as the unknown creature used both its small, clawed hands and its snout to search for food within the pile.

Eventually the creature finished its search and made its way back to the water. George made his way towards the location where he watched the creature rummage for scraps and took notice of a trail of webbed tracks in the loose sand and dirt that led both to and from the cold water...

The second most often talked about sighting regarding the Alligator took place in 1915. Three men by the names of Charles Flood, Green Hicks, and Donald Macrae were in the area of Hope, British Columbia, when they came upon a small mud lake. Within this lake the men noticed what appeared to be small, almost baby looking alligators that were only a few inches long swimming around in the mud. What was striking to all three men were that the creatures were completely black in color and unlike anything they had ever seen before...

In 1973, the third most famous sighting of the Canadian Alligator would take place on Pitt Lake. It is this specific lake that gives the creature is second (arguably more famous) name, the Pitt Lake Lizard. On June 3rd, married couple Warren and Sharon Scott watched in disbelief as a large number of creatures that resembled huge reptiles swam slowly through the water while closely grouped together..." [6]

According to the book *Strange Creatures Seldom Seen* by John Warms:

"One man from Pukatawagan whom I spoke with killed a brown, four-foot crocodile-like creature in a log trap..." [7]

Pukatawagan is a town in the far West of Manitoba, Canada. In a lake where the Winnipeg river enters Lake Winnipeg:

"There, a few feet below the surface, was a larger version of the "mud puppies" they would sometimes snag with their fishhooks in the river– a small lizard-like black salamander. This one, with its large bulbous head, was identical in all aspects except in size, for it was about ten feet long. It swam like a crocodile using its legs, but its movement also had a snake-like appearance..." [8]

Shane writes: Monsters ARE real! 2 real-life animals that would qualify as being monsters, are the Japanese and Chinese giant salamanders. They really

exist and grow up to 5-6 feet (not inches, feet!), in length.

The Japanese giant salamander (*Andrias japonicus*), is a mottled black and brown colour, with wart-covered-skin and is fully aquatic, living in the fast-flowing streams of Japan. It grows up to 5 feet in length, and is the 3rd largest salamander in the world.

There is a curious legend of an impish, reptile-like being, in Japan, called the Kappa. They look for all the world like Teenage Mutant Ninja Turtles! The Kappa is sometimes depicted as being bipedal and sometimes quadrupedal. According to Japanese legends, the Kappa had a propensity for abducting people and playing crude, sometimes cruel jokes on humans.

The Japanese people clearly morphed something into a living legend, that represented Japanese humour and predilection for wrestling (we know how they love Sumo-wrestling). The Japanese clearly had fun with this legend. Could there be any truth behind these ribald tales? Quite possibly, the Japanese giant salamander was the original model for the legends of the strange Kappa beings.

The Japanese giant salamander with it's imposing size, bizarre appearance and intimidating manner of darting out of nowhere and snatching prey, more than likely inspired fear, respect and superstition in the ancient Japanese people, long ago. That being said, a real animal, in this case an animal known to science, but, a real animal nevertheless, was probably the original template that formed the Kappa legends.

These legends originated in Japan, but, are famous worldwide.

Could there be any truth to other legends, sightings or traditions of giant salamanders worldwide? Yes!

As mentioned earlier, the American state of California has had a long-standing history of reports of giant salamanders, among old-timers (miners, trappers, hunters), in and around the Trinity Alps region of northern California and other parts of northern California.

In 1960, animal handler Van Harden from Pioneer, CA, reported that he had actually hooked 1 of 8 giant salamanders that he had seen in a Trinity Alps' wilderness lake. Harden claimed to have actually measured the hooked salamander at 8 feet, 4 inches (!), before he released it back into the lake. Catch and release.

One of the earliest reports of a possible giant salamander in northern California, occurred in 1891, when the "head" of a gigantic lizard was seen in the Sacramento River, near Woodland, California. The poor animal was even shot at, as it attempted to climb out of the river and on to the shore. It was last seen swimming as fast as it could, away from there! Can you blame it? Now, who's the bigger monster, man or beast?

Similar reports of giant amphibians come from Canada, where they are often referred to as "black logs", or "black alligators".

Assuming that these reports are genuine and reliable, what kind of amphibian could be responsible for the unique reports emanating from the Trinity Alps, Sacramento River and Canada?

North America has many amphibian species, with 198 salamander species living in the United States and 21 in Canada.

In the Ozark and Appalachian Mountains of the United States, there lives a giant salamander called the Hellbender (*Cryptobranchus alleganiensis*). It is the largest salamander in North America, reaching lengths of up to 29 inches and lives in fast-flowing streams, like it's Asian relatives, the Japanese and Chinese giant salamanders. It too, is a member of the family Cryptobranchidae, just like the Chinese and Japanese giant salamanders. Could certain Hellbenders attain larger lengths? That would be one hell-of-a Hellbender! Pun intended. Larger Hellbenders are a distinct possibility. There could also be unknown populations of large Hellbenders in certain areas, possibly accounting for giant salamander sightings and reports.

One other potential candidate for mystery giant salamander reports, is the Pacific giant salamander (*Dicamptodon tenebrosus*). They are endemic to California, the Pacific-North-West of the United States and Canada. They live in cold, mountain streams and lakes. Could the occasional acromegalic individual, account for these reports? Maybe. The Pacific giant salamander usually only reaches 13.4 inches in length. Acromegalic individuals could theoretically be larger. Acromegaly is a condition in which the body produces an excessive amount of growth hormone, due to an abnormality in the Pituitary Gland. The excessive amount of growth hormone, in turn, produces gigantism.

An example in humans, is 7 foot, 2 inch actor Richard Kiel - "Jaws"- of the James Bond movies "The Spy Who Loved Me" and "Moonraker". Kiel also played Dr. Loveless' sidekick "Voltaire", in The Wild Wild West series. Also, Kiel, to the delight of cryptid fans worldwide, played two creatures of sorts, in the Kolchak The Night Stalker series- a giant Amerindian shaman spirit, called a Diablero and a Spanish Moss monster, from Creole legend, called Pere Malfait (father of mischief).

Isolated populations of animals can result in gigantism. But, the more large individuals you have and the larger the geographic area they cover, acromegaly and isolation become more untenable. The cases with groups of large salamanders, or cases featuring unusual colouration are much harder to explain away.

Since North America already harbours a member of the family Cryptobranchidae (the same family that contains the giant Asian species), there could very possibly be an uncatalogued giant salamander existing there, closely related to the extant, giant Asian salamanders. Fauna and flora wise, there are ancient connections between China and North America. Take alligators, for example. Alligators are only found in 3 places, in the entire world, the United States, Mexico and China. This zoo-geographical relationship, shared between

North America and China, is even more circumstantial evidence, for the possible existence, of a giant salamander, in the cold-water lakes and streams of California and Canada.

REFERENCES

1. Loren Coleman. Promises of Giants. Fortean Times # 103. October 1997 p.43
2. George S.Myers. Asiatic Giant Salamander Caught in the Sacramento River. Copeia. June 1951 pp 179-180.
3. Loren Coleman Ibid
4. Loren Coleman Ibid
5. Thomas Rodgers. Report of Giant Salamander in California. Copeia. September 1962 pp 646-647
6. Cryptid Profile: Canadian Alligator (AKA Pitt Lake Lizard) -The Pine Barrens Institute
7. John Warms. Strange Creatures Seldom Seen. P. 168
8. John Warms. Ibid. p. 170.

Mystery giant salamanders of the World Part Two — Europe and Asia

Shane Lea and Richard Muirhead

Continuing our essay on mystery giant salamanders of the world, we now focus on cases in Europe and Asia. There is a strand of opinion that the Loch Ness Monster, far from being a plesiosaur, is in fact a giant salamander. The Loch Ness monster, it has been suggested, could be a plesiosaur, a giant eel, a seal, an otter, or the branch of a tree.

There is a blog by Steve Plambeck, titled *New Evidence for Giant Salamanders in Post-Glacial Europe*, dated October 6th, 2017, in which he attempts to make the case for the Loch Ness monster being a giant salamander originating in what is now Turkey and migrating to the loch via Doggerland, now submerged beneath the North Sea. The author states:

> "...So now to the question: How could a theoretical Loch Ness Giant Salamander, that being the thesis of this blog, even stand a chance unless there was a surviving European Giant Salamander first? Pre-ice age salamanders don't count, because Loch Ness didn't exist until after the last glaciation. Europe today is replete with small newts and salamanders of the modern suborder *Salamandroidea*, and then there's the olms, the ancestors of which came racing into Europe before the last glaciation had even finished melting. All of these prove post-glacial Europe is an amenable ecosystem for small, modern salamanders and all manner of other amphibians, but what about those primitive Giant Salamanders that had thrived in Europe before the ice came? I find it a bit surprising that a family that endured the K-Pg mass extinction and flourished another 65 million years couldn't survive a 98,000 year vacation to the nearby Mediterranean, like the rest of the fauna. There could be reasons of course, but Occam's razor leads us to consider a different possibility..." [1]

This "different possibility" is that giant salamanders migrated from Göbekli Tepe in Turkey to Loch Ness via Doggerland. Plambeck came across an archaeological site there where a carving of a giant salamander on stone is to be found.

"Two of the more bizarre Loch Ness sightings happened in 1888 and 1933. In '88, Alexander MacDonald reported what he described as a salamander-like creature with "large stubby legs" surfacing in the water. And in '33, strangest of all, a man named George Spicer and his wife said they witnessed a 25-feet-long creature with a wavy neck cross the road right in front of their car, heading for the loch..." [2]

There is an interesting comment by 'Anonymous' on March 21st, 2014, at the end of the blog 'An Niseag vs Nessie—Folklore vs Science' of March 10th, 2014, by Steve Plambeck:

"Special bonus for salamander fans:
Early in the Second World War, Ian [MacDonald] had been on an exercise in the area with the Cameron Highlanders when, going along the shores of Loch Ness, a disturbance was spotted. They all piled out of the truck and, armed with their spying binoculars, they watched "the beast" for a full twenty minutes. When Hamish [MacDougall] as a little lad was not behaving to Granny's [Annie MacDonald nee Galbraith's] liking, she would threaten "I will throw you to the salamander" – her name for "the monster" ." Peter R. English "A bridge to the past: an oral history of families of Upper Glen Urquhart" (Inverness: Speedprint, 2009)

In July 2022, it was announced that the fossil of a giant salamander had been found on the isle of Skye, Scotland.

"The research team analysed 166-million-year-old fossils of a type of animal called Marmorerpeton, found in Middle Jurassic rocks on the Isle of Skye. They found that it has several key salamander traits, but is not part of the modern group of salamanders. Their results are reported in Proceedings of the National Academy of Sciences (PNAS). The specimen is believed to be the oldest salamander fossil found in Europe." (Oldest European salamander fossil, discovered in Scotland, informs amphibian origins. July 12th 2022. www.ucl.ac.uk)

Roy Mackal in his book *The Monsters of Loch Ness* points out that the crest of a newt could explain the humps of the Loch Ness Monster, though it would have to be a pretty huge newt! According to Karl Shuker in his book *Here's Nessie!*, referring to Professor Roy Mackal's book:

"In it Prof.Mackal meticulously examined every reasonable zoological identity, and concluded that the most plausible Nessie candidate was a species of giant newt or salamander-like amphibian, which in his view would account for 88 per cent of the LNM characteristics on file (as opposed to 78 per cent for a species of eel, 69 per cent for a plesiosaur,

59 per cent for a mollusc, 56 per cent for a seal, and 47 per cent for a sea-cow).

"There have been other mystery giant salamander sightings in the United Kingdom apart from the case of Loch Ness. In a book titled `In The West Country` by Francis A. Knight (1896) mention is made of a highly venomous "triton" in the Somerset Levels as follows: " Slender newts, too, swarm in the still water, and great black tritons, the terror of the moorfolk, in whose eyes even the viper is hardly more venomous". (page 186.)

In *Flying Snake* Volume 5 no. 13, May 2018, I commented:

"Now this is very interesting because long ago the triton was thought to be a kind of mermaid or large sea snail, according to Wikipedia. Or perhaps a salamander. Francis Knight specifically says they were distinguished from newts, or perhaps they were newts but of a darker colouration? The Somerset Moors are in north Somerset and the Somerset Levels not far away.

EDITOR'S NOTE: Triton is an archaic word for newt, and the word can still be found as part of Latin binomial names. My interpretation of this, particularly exciting, record is that at one time, at least larger, black coloured newts, possibly melanistic, great crested newt, or possibly something even more exciting could be found in the Somerset levels..

According to the *Surrey Advertiser* of August 8th, 1908, a fire salamander was discovered in Surrey:

"The Recent Discovery of a Lizard:
Mr F.H.Elsley writes: The lizard recently found on the Hog`s Back has been sent to the Educational Museum at Haslemere. It proves to be from Central and Southern Europe...How it came to be found on the Hog`s Back is certainly a mystery." (see *Flying Snake* Volume 4 Number 12 September 2017 pp 60-61)

Jumping forwards to 2015, according to an article in *Animals & Men* Issue 53, May 2015, by Steve Baxter, he was exploring a pond in a quarry not far from his home in the village of West Mickley near Stocksfield in Northumberland. He lifted up a large flat log " to find what I `think` were European salamanders…"

Presumably he meant fire salamanders, see image on the following page.

Elsewhere in Europe, there is the case of the Tatzelwurm, a salamander-like cryptid said to inhabit the Bavarian, Swiss and Austrian alps, said to be aggressive and

Animals & Men

A COLONY OF FIRE SALAMANDERS IN
NORTHUMBERLAND?
An interview with Wes Sullivan; Sea Serpents;
Hunting the Blue Devil; Butterfly invaders; the
science of Owlman and more

The Journal of the Centre for Fortean Zoology; Issue 53 - May 2015

venomous, compared by Bernard Heuvelmans, the father of cryptozoology, to the Gila Monster of the south-west United States, (Arizona and New Mexico.)

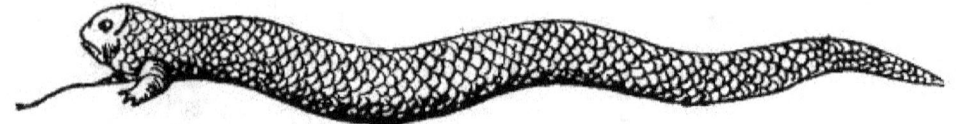

The Tatzelwurm is 60cm to 90cm long, and looks like a worm with stumpy short feet. So, vaguely salamander-like in appearance. In 1861, Friedrich von Tschundi in his work on Alpine fauna *Das Theirleben der Alpenwelt* wrote:

> "In the Bernese Oberland and the Jura the belief is widespread that there exists a sort of `cave-worm` which is thick, 30cm to 90cm long, and has two short legs; it appears at the approach of storms after a long dry spell."
> " (Bernard Heuvelmans *On The Track of Unknown Animals* p.11.)

From the point of view of the Tatzelwurm being a mystery giant salamander, it is interesting that the Tatzelwurm is said to appear on the eve of storms, because the Hong Kong mystery giant salamander *Megalobatrachus sligoi* (now known as *Andrias sligoi*) appeared in a drain after a "particularly violent" (Wikipedia—South China giant salamander) rain storm in April 1920.

In the same book just quoted from Heuvelmans says:

> "There can be no doubt that it exists. An enquiry made in the 1930s by Dr Gerhard Venzmer and engineer Hans Flucher brought to light the evidence of some 60 witnesses. They all agreed that it was between 30cm and 90 cm long and more or less cylindrical in shape, the tail end of the body ending rather abruptly. It was brownish on the back and lighter underneath. It had a short and thick tail; there was no narrowing at the neck; and its heavy and blunt nosed head had large, round eyes. Its legs were slender and small; and some even maintained that it had no hind-legs at all. Some, but not all, said it had scales, but in any case it hissed like a snake." (Heuvelmans, ibid, pp 11-12.)

Heuvelmans states categorically (p.17) thus:

> "Whether the Tatzelwurm is a European species of Heloderma , an enormous skink, a giant salamander or some completely new animal, there is little doubt that it exists…"

Dr Karl Shuker (personal communication , August 11th 2023) believes the Tatzelwurm is a kind of reptile because it is covered in scales. Rosenrot posted the following comment on the Unexplained Mysteries.com Forum on June 3rd, 2007:

> "The theory of the Tatzelwurm being a giant salamander is rather interesting. If I remember correctly giant salamanders can live in rather cold climates which would explain the "Tatzelwurm" being seen all over Europe. But I don't know how vicious giant salamanders are...

Didn't the Tatzelwurm actually try to attack a few people?

Now turning to Asia: The most well known mystery giant salamander in Asia is *Megalobatrachus sligoi*, or *Andreas sligoi* as it is now known. In April 1920, after a heavy shower of rain, a giant salamander was discovered in a depression in the ground at the Hong Kong Zoological and Botanical Gardens. During this particularly violent storm, a large drain pipe in the Gardens burst, carving a large depression into the land that the escaped salamander was washed into. It was captured and kept in a large circular basin, where it was fed daily with live tadpoles and occasionally beef . (Wikipedia—South China Giant Salamander.)

Wikipedia. Holotype of Andrias sligoi. (BMNH 1945. 11.7.1.): dorsal view.

Wikipedia goes on to say:

> "The captured salamander was later seen by George Ulick Browne, the then-Marquess of Sligo, as he was touring the area. Browne persuaded the then-governor of Hong Kong, Reginald Edward Stubbs, to present the salamander to the Zoological Society of London. Upon receiving the individual, Boulenger found it to be physically distinct from "Megalobatrachus maximus" (the former species into which his father, George Albert Boulenger, lumped both the Japanese giant salamander and Chinese giant salamander) and it thus likely represented a new species. During Boulenger's description, he named the species M. sligoi in honor of Browne's title…"

It could be argued, as Wikipedia does, that the origin of this salamander was southern Guangdong province, after all, English language Hong Kong newspapers in the first decade of the Twentieth century described "fish with legs" in the Pearl River estuary which divides Macau on the west side with Hong Kong on the east.

However there is a precedent for animals to live in Hong Kong, whilst not being mentioned in text books of Hong Kong zoology; I am referring to the Axis deer and the dhole. Also there are adequate pools and streams on Hong Kong island and in the New Territories, which could have contained giant salamanders, especially in the 1920s in the then British colony of Hong Kong, before the days of urban encroachment upon the countryside. Several years ago I did some further research into the origins of Sligo's salamander using the South *China Morning Post* online newspaper archive and I came

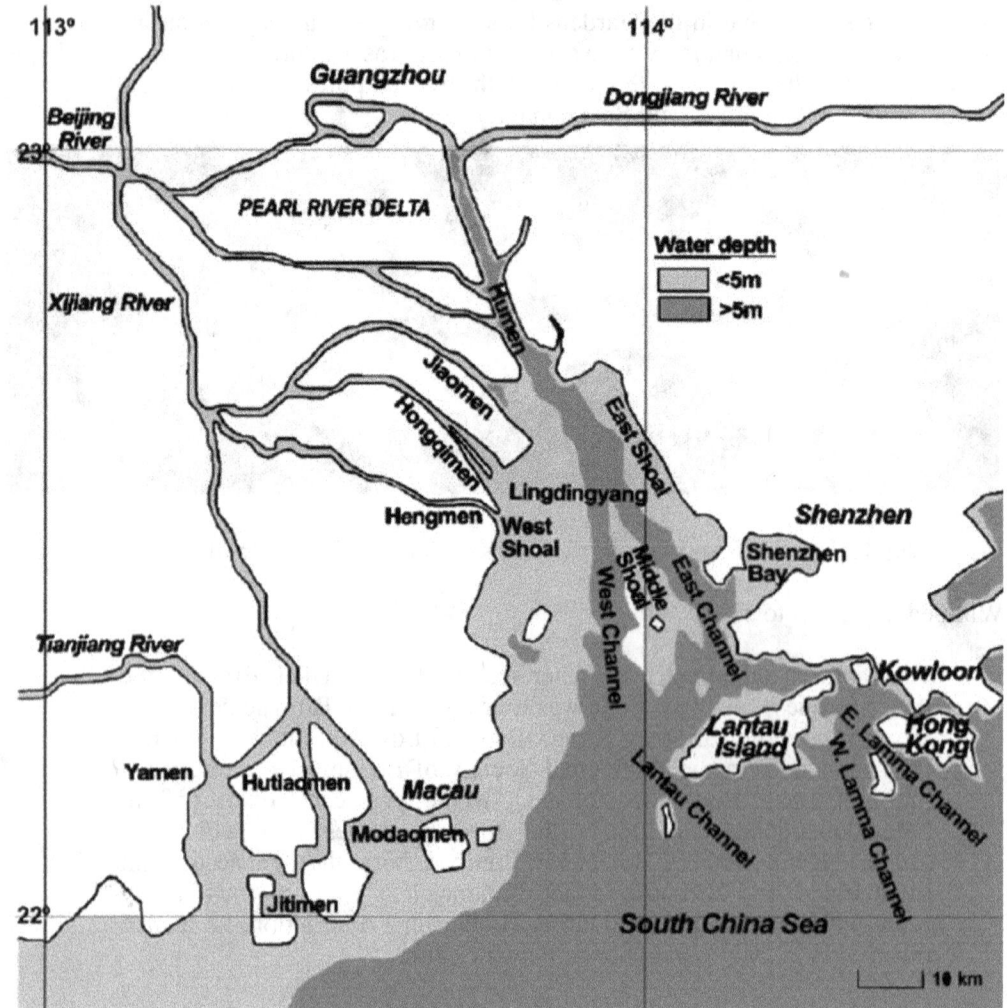

across the fascinating story of Mr Crook, who in a letter to the S.C.M.P stated that in 1909 he:

"...took two salamanders and put them in a stream on the Peak which flowed from near Mountain Lodge and emptied itself into the sea near Pokfulam. Years afterwards I went to the Zoological Gardens here in London to see a giant salamander which was on exhibit there. The announcement said it had been obtained in Hong Kong. *I often wondered if it was one of the grand-children of the ones I had let loose in the stream on the Peak* (Emphasis my own.) [3]

I did some further research and found out that there was a conduit running around to the northern part of Hong Kong island to the vicinity of the Hong Kong Zoological and Botanical Gardens. Maybe the salamander travelled along this conduit to reach the Gardens? See *Animals & Men* issue #53, May 2015, for the complete story. Mr Crook`s letter to the *South China Morning Post* is dated July 3rd, 1953.

The *Hong Kong Police Magazine* for March 1956, reported on another probable giant salamander, this time in the Wanchai district of northern Hong Kong island. A police officer saw an animal that emerged from a drain. It was about eighteen inches long, had four legs, a spade head , a tail and a mouth full of saw-edged teeth. The article says:

"It was subsequently removed to the Hong Kong University, where it was later identified as a `Giant Salamander`. Many of these are evidently imported into the Colony for sale in the local markets. As a matter of interest, they are perfectly harmless and do not possess saw-edged teeth…" [4]

According to Dale Drinnon`s Frontiers of Zoology blog `The Giant Salamander Theory at Loch Ness` of June 7th, 2011, referring to Rupert T. Gould`s book *Loch Ness Monster*:

"The suggestion that X might be some species of giant salamander" which is "indigenous to Loch Ness and its rivers" was made by Lt-Col. W. H. Lane, Glenmoriston, in a letter to the *Inverness Courier* (10.x.33). While pointing out that the largest living variety is a native of Japan (Although specimens have been obtained in China), he stated that he had shot what he believed to be a creature of the kind in the Chin Hills district of Burma [5]

Shane adds: Continuing with the line of thinking from part one, there is the distinct possibility of undiscovered, giant salamanders in California, U.S.A. and Canada. The California reports were covered in part one of this article, so, next up are some Canadian reports that allude to giant salamanders.

There have been unexplained reports of "black logs" and "black alligators" throughout Canada, dating back many years. Our interpretation is that what is actually being witnessed are in fact giant salamanders.

Many of the Canadian giant salamander reports fall under the umbrella term of Lake Monsters.

In Cryptozoology, every distinct subject usually encompasses a wide variety of animal reports and/or phenomena. Rarely is a subject so cut and dried. It is the Cryptozoologist's mission to determine, if indeed an unknown animal is part of the dossier of reports, and to determine what that animal most likely is within the framework of zoological classification and to determine the most likely way, based on a study of the animal, to prove its existence, thereby ensuring the survival of it's species and the protection of it's environment.

Having said that, when examining lake monster reports, one is subject to a complex variety of boat wakes, wave patterns, atmospheric phenomena, misidentified known animals, possible unknown, unclassified animals and sometimes, even more anomalous phenomena. So, it gets complex, sorting it all out.

One aspect that can explain some Canadian and American reports, is Sturgeon. No doubt, these account for many lake monster reports in the United States and Canada. Sturgeon are very big, impressive fish, that can surface looking like a "prehistoric monster", because, they are indeed prehistoric fish. Their appearance is startling and many times are not seen for what they are, because only their jagged back is seen, looking very bizarre. Lake Sturgeon (*Acipenser fulvescens*), grow up to 8 feet and are a fresh-water species, whose fossils date back to the Pleistocene Epoch. Atlantic Sturgeon (*Acipenser oxyrinchus oxyrinchus*), a marine species, grow up to 15 feet and could be seen while swimming upstream to spawn, in eastern Canada, or America. It is considered a living fossil.

Atlantic sturgeon. *Acipenser oxyrhynchus*. Wikipedia.

So, eliminating Sturgeon confusion from the "black alligator" dossier, one is left with a core of unexplained reports, alluding to a very sizeable animal living in Canada and California.

Escaped/abandoned pets have also to be considered. A cute, little, family pet of 6 inches can quickly grow to a 6-foot toothy menace! Then, to give the animal a chance, people release unwanted pets into the wild, not realising the disruption they cause. In the case of the "black alligators", escaped/abandoned pets could not survive Canadian winters and the sheer number and description of the reports over long periods of time also eliminates the "pet" explanation, so beloved by cryptozoological critics, cynics and Philistines.

Cryptozoologists do indeed debunk certain reports or subjects, when debunking is warranted, without any acknowledgement from cryptozoological critics, cynics and Philistines. So, in essence, Cryptozoology is self-correcting.

In the case of the possibility of giant salamanders in Canada and California, once the other possibilities have been eliminated, one is still left with the fascinating prospect of an unknown animal, which, of course, what Cryptozoology is all about, those possibilities.....

Canadian black alligator reports date back over 100 years.

One George Goudreau, witnessed an alligator-like animal emerge from Lake Kootenay, British Columbia, on 10 October, 1900. The "alligator" was seen from a distance of only 12 feet away, as it emerged from the lake onto land and moved toward a pile of rotten compost on shore, searching for food. The creature, about 3 feet long, black and green in colour, was observed to have small, clawed feet, 10 inches long and used it's claws and snout to move around the pile of rubbish, looking for tasty morsels. Eventually the creature went back into the lake, leaving behind a trail of webbed feet on the sandy shoreline.

Later on, a similar animal, 15 feet long, was seen by A.C.D. Robertson in Kootenay Lake. He witnessed a large, scaly body that raised out of the lake to seize fish that had been tied to a stake.

In Kootenay Lake, we have reports of different sized creatures, thereby eliminating the escaped pet theory, that was untenable to begin with, since pet alligators can't survive Canadian winters, to grow to some of the sizes reported. In contrast, the Asian giant salamanders are found in cold-water streams and rivers.

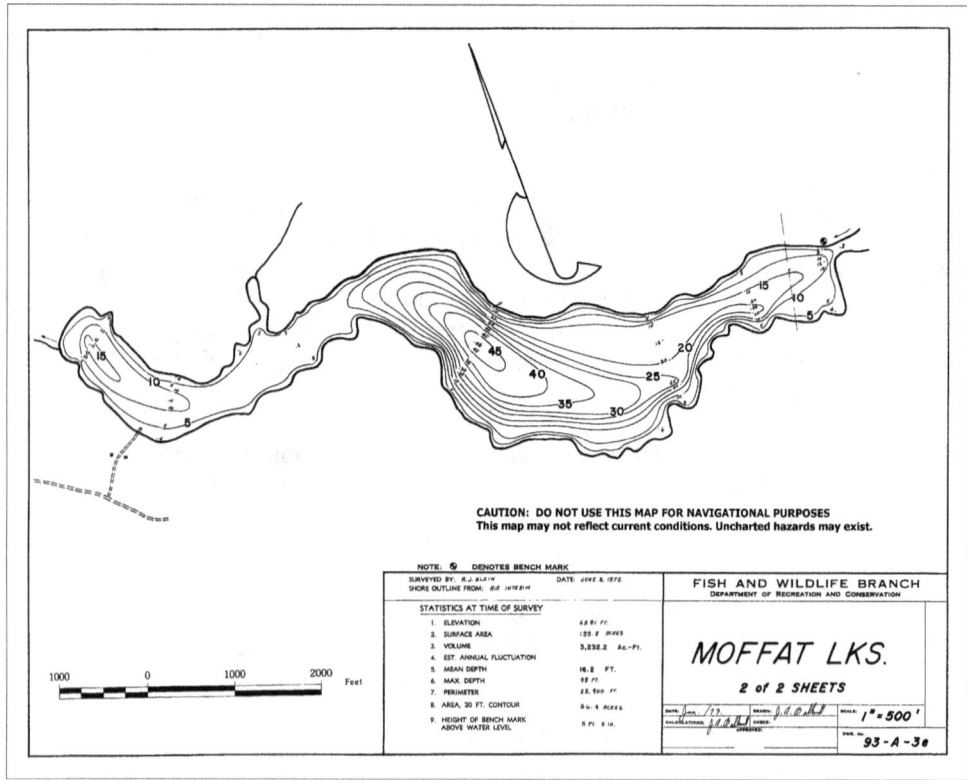

In 1915, small black alligators were seen in a small mud Lake, in the Hope Valley area of British Columbia, by Charles Flood, Green Hicks and Donald Macrae. The 3 men tried unsuccessfully to capture some of the small, slippery, completely black, "alligators", which were only a few inches long.

Harvey MacRae, stated that his father had seen a big, round, black, 15-20 foot long "tree trunk", lying low in the lake water of Moffat Lake, Quebec. Onshore of Lake Moffat, a big, burnt log was seen stretched across a foot-path near the lake. The "log" was seen to rise up and scuttle off toward the lake!

A recent sighting of a possible giant salamander from Pitt Lake, British Columbia, dates from 2002. In 2002, Pitt River Lodge owner Dan Gerak saw a creature that was over 5 feet long, on 2 occasions, that had smooth, dark, black skin and looked like an over-sized salamander. Yes, over 5 feet long! This is on par with the size of the Asiatic giant salamanders. But, we're still talking about Canada, right? Right!

Pitt Lake British Columbia . Wikipedia

Similar creatures have been seen in Cultus Lake, Chilliwack Lake, Nitinak Lake and the Fraser River, in British Columbia. Other reports emanate from lakes in Quebec, as well.

As discussed in part one of this paper, North America shares many affinities to China, relating to flora and fauna. An undiscovered population of an unknown species of giant salamander, in North America, similar in morphological size and appearance, ecologically inhabiting a similar habitat, with similar behavioural traits, to the giant Asian salamanders, would not be out of the question. The discovery of such an unknown animal in North America, is well within the realm of possibility.

What about the possibility of giant salamanders in other parts of the world? Well, Richard discussed many of those intriguing subjects in this article.

That having been said, one very interesting prospect has to do with giant salamander reports from Lake Titicaca, Bolivia and giant salamander reports from elsewhere in South America.

In the region of the headwaters of the Amazon River, in South America, a giant salamander has been reported since the 1950's, by various Amerindian tribes and timber workers. The animal is called Axolotlis, and has been described both as a slimy dog skittering through the forest and as a crocodile (caiman?) sized salamander, reaching lengths of 15-20 feet. The name of this cryptid is very similar to the Axolotl salamander from Mexico.

Giant salamanders with frog faces and bulging eyes, have been seen in Lake Titicaca, Bolivia. There are also reports there of giant tadpoles. These reports could possibly refer to a new population of Axolotl salamanders. The "frog faces" and "bulging eyes" are a dead-ringer for an Axolotl. Even a description such as "giant tadpole", could describe an Axolotl amphibian.

Axolotls are neotenous amphibians, meaning that they retain their large, feathery gills into adulthood and remain in an aquatic environment, rather than adopting a more typical, terrestrial existence. Axolotls are critically endangered where they are currently found, in 1 lake only, in the proximity of Mexico City, Mexico. There are no species of giant Axolotls currently known to science. They reach a 1 foot maximum length. A new, distinct population may have a different size range, or, it could be the case that the giant salamanders reported from Lake Titicaca are Axolotls similar-sized to the Mexican species, yet, possibly, a new species of Axolotl.

Axolotls have amazing regenerative powers. They can even grow back and replace lost limbs! Amazing! We have much to learn from these awesome amphibians. If new populations of Axolotl were found to exist, it would be wonderful news for zoology and especially, for the conservation and future of this adorable, highly endangered, amazing amphibian species.

Axolotl

The same holds true for any of the giant salamanders discussed in this article. We are talking about "real" animals here and the zoological discovery of any of them would be sensational, with the ultimate goal being along with their scientific and worldwide acknowledgement, the protection of the new animal species and it's habitat. This would ensure the protection of the cryptid and all other plant and animal species sharing it's ecological habitat.

In 1984, the world population was 4 billion. Now, the world population is 8 billion. Double! As man continues to encroach on all living things, destroy habitats, pollute the environment, exterminate for "bush meat", or "traditional medicine", wage senseless, "war pig" wars, on our own kind and dumb-down the mind, through over-reliance on machines, let us hope for the animals' sake and ours, that a few far-sighted Cryptozoologists will continue to stay On The Track Of Unknown Animals.

The Great Days Of Zoology Are *Not* Done!

REFERENCES

1. Steve Plambeck New Evidence for Giant Salamanders in Post-Glacial Europe. October 6th 2017.
2. Rob Schwarz. Multiple Large Unidentified Objects Spotted On Shores Of Loch Ness - Stranger Dimensions. October 2nd 2020.
3. Letter from R.Muirhead to Animals and Men Issue 53 May 2015 pp 75-76.
4. Letter from R.Muirhead to Animals and Men Issue 54 2015 p.71
5. The Giant Salamander Theory at Loch Ness. Frontiers of Zoology. June 7th 2011.

A different edit of this two part article appears in Flying Snake # 26 and # 27 and appears courtesy of Richard Muirhead

Index to *Flying Snake* #11-15

Richard Muirhead

Giant hail in China   Sep-17
Giant lizard, Himalayas   Jul-19
Giant snail   Sep-17
Giant snake hoax, Cambodia   Jan-19
Giant turtle, Persian Gulf   Jan-19
Giant wasp, Scotland   May-18
Giant worm, Macclesfield   Jan-19
Ginger rabbits   Jan-19
Giraffes and creationism   Jan-19
Gorilla in California Sep-17
Greek coin, found in dress   May-18
Greek statue, ancient, New York   May-18
Hairy elephant, Sumatra   Sep-17
Half cat half rabbit   Sep-17
Horse, strange   Sep-16
Human skeletons in trees   Sep-17
Huge cow, Australia Jan-19
Huge eggs, New Mexico   Sep-17
Huge hail, Sydney   Jan-19
Huge frogs   Sep-16
Huge ocean eel   Sep-16
Huge otter, Scotland Jan-19
Huge python, Hong Kong   Jan-19
Huge ray, Pacific   Jan-19
Huge tadpole Jan-19
Human skeletons in trees   Sep-17
Hyena in India   Sep-17
Hyena-like animal   Sep-16
Insect cryptid, Arizona   May-18
Jackal in Ireland   Jan-19
King Henry VIII`s sea monster   Sep-17
Large eel, Lake Bala Jan-19
Large Ox   Sep-16
Lepidopterophobia   Sep-17
Lions, France Sep-16
Lions of Valencia, Spain   Sep-17
Living dragon, Hungary   May-18
Living mastodon   Sep-16
Lizard in a block of coal   Sep-16
Loch Ness Monster   Sep-16
Loch Ness Monster sculpture   Jan-19
Longest lasting rainbow, Taiwan   May-18
Loyal dog, China   Jan-19
Lynx, Cheshire   Sep-16
Lynx re-introduction Jul-19
Madstone   Sep-16
Man-eating snakes, Fiji   May-18
Man persecuting toads   Jul-19
Manta ray, China   Sep-16

CFZ, ANNUAL REPORT 2022

Dear Friends,

Every year for the last 28 years I have sat down in the week before Christmas and typed or dictated my annual report on how and what the CFZ has been doing during the previous year.

Claims that I title the report after the series of albums by Throbbing Gristle are something I will neither confirm nor deny.

Things have changed at the Centre for Fortean Zoology many times over the intervening 28 years. Sadly, since about 2017, my profile both within the Cryptozoology community and indeed within the CFZ itself, has been fairly low-key. In those years, I nursed Mother who finally died of dementia at the end of 2019 and I then nursed my late wife Corinna through what turned out to be her final illness before she died of cancer in 2020. I have had a whole string of health problems of my own and as anybody who reads my regular bulletins will know - it has taken me several years to get back to fighting strength after all of that. However, I am pretty much back to speed now or as much as I am ever going to be, and it is good to be able to report that the CFZ is beginning to expand again after all those enforced years in the doldrums.

2022 has been a year which has seen the launch of several new projects and a particularly successful expedition to Sumatra.

Odette Tchernine
You may remember that a couple of years ago, Richard Freeman and I were approached by a man whom Richard had once got chatting to on a train. As a result of this conversation, he remembered that Richard was working with the CFZ and was an avid student of cryptozoology. Apparently, he was at an auction and, amongst other things, bought some documents including an unpublished book by the late Odette Tchernine. We purchased these items from him for a couple of hundred quid and I immediately set about the ever-resourceful Guin Palmer (referred to as Miss Guinevere in 'On the Track') to the task of seeing if she could find Odette's surviving relatives.

What I was not expecting, was that Guin would become totally fascinated by Odette Tchernine and her legacy and that she had bequeathed enormous amounts of supplementary information, articles, plays, and poems all written 80 years ago. Moreover, Guin also made contact with a lady, also living in Devon. who had purchased a considerable number of documents from Odette Tchernine's estate and who was happy to collaborate with her on what would soon become quite a major project.

I am certain you will remember that last year Louis Rozier built us a magnificent new website. By the late spring, we had amassed so much information and documents relating to Odette and her legacy, that it was decided we would present them to the world in a similarly all-singing, all-dancing website, designed by Louis, and hopefully paid for by the Arts Council. In addition, Guin (who is somewhat of a grammar Nazi) is editing and putting together Odette's unpublished books about Man Beasts, working together with Richard Freeman on the footnotes and endnotes. We confidently expect it to be a major piece of scholarly work as was 'The Soviet Sasquatch', the long-lost book by Boris Porshnev which we finally published in August of 2021, after years of work by Dr Chris Clark and Lars Thomas.

Finally, Guin has decided to write a biography of Odette Tchernine, featuring all the disparate information that she has collected over the last year and is continuing to collect. The cumulative source material will also appear on Louis' all-singing, all-dancing website.

I would like to publicly thank Guin for all her hard work on this project and look forward greatly to seeing it progress over the forthcoming 12 months.

By the way, just in case you have ever wondered why I refer to Guin as Miss Guinevere and Maxine as Miss Maxine, and do the same for other ladies working within the CFZ, I refer you to an obscure arch rock band from Laurel Canyon in the late 1960s. They were called the GTOs and consisted of a bunch of female acquaintances of Frank Zappa (including his babysitter). He named them all with the prefix "Miss', including Miss Pamela, Miss Christine and Miss Mercy, and the whole thing was referenced in one of John Waters' movies. I find it more than slightly amusing that I first referred to Maxine as 'Miss Maxine' when I made the Allman movie back in 1998/99. I have continued this ever since mostly because it amuses me.

CFZ People

This year has seen some notable comings and goings in CFZ personnel. Sarah, our housekeeper for the last four or five years, left for personal reasons and has been replaced by not one, but two delightful ladies, Judy and Shelly. They spoil me massively, mother me, and do a fantastic job. The house looks nicer than it has since my parents died, and I enjoy their visits immensely.

As you may remember, apart from being my much-loved stepdaughter, Olivia was also my secretary for quite a few years. She left when she got a full-time job, in the summer of 2021. Various people, including Louis, took over secretarial duties in the meantime, but none of them were really fit for purpose but now I am incredibly pleased to say that this autumn, I acquired the services of a brand-new sparkling secretary called Karen who is an absolute gem.

Robin Pyatt Bellamy was our Canadian representative for many years, until she had to retire on health grounds last year. I would like to stress "health grounds" are not a euphemism and she is very seriously ill. She and I still chat intermittently online and I am very fond of her. However, I am very pleased to say that she has been replaced by David Scott and his family, including his daughter Carol, whom many of you will remember as a recent guest on 'On the Track', where she was talking about her new book about the mystery animals of Oceania. Whilst on the subject of regional representatives, Katy Elizabeth, best known as a researcher on Lake Champlain, has taken over as the United States representative.

I am also particularly pleased to welcome Saarthak Halbar as our representative for India, and Jakub Roček as our representative in the Czech Republic. The latter is the head of the Ursus Arctos Project which is studying brown bears in Slovakia and the persistent stories of brown bears in the Czech Republic where they are supposed to have extirpated. Dally Sandradiputra is now our representative in Indonesia and I hope to meet him face to face when he visits the UK next year for the Fortean film festival. Finally, Trần Văn Xám is our representative in Vietnam.

I feel that I should probably have a Bosun's whistle to pipe all these people aboard, but I looked on Ebay and they were remarkably expensive, so I haven't got one.

Lake Champlain Project
Katy Elizabeth has been a friend of the CFZ for some years now, ever since Richard and I rode separately into battle to protect her from the unfortunate sexism, which is so rife in various parts of International Cryptozoology. She is a well-known researcher on Lake Champlain and in September she managed to film some extraordinary sonar footage that appears to be of an animate object in 40 feet of water; the creature, if indeed it is a creature, being about 20 feet in length. I had a long discussion with her last night, (19th December) and I am happy to say that she is going to be publishing a copy of her findings in a couple of years in a specially constructed corner of the CFZ website. I look forward to seeing how this progresses.

Sumatra 2022 Expedition
As you probably know, we sent our sixth expedition to Sumatra in September and they came back to dear old Blighty after nearly a month in the jungle. It was probably our most successful expedition yet, at least since 2009; the one where our guide Sahar Dimus, and expedition member Dave Archer, both saw the orang-pendek in a tree and where hair samples were obtained. For those of you who don't know, Danish zoologist Lars Thomas, who runs the CFZ lab from his home in Copenhagen, and works closely with Dr Pen Gilbert at the Copenhagen University, declared that the hair samples were close to those of an orangutan but distinct enough to be a different species. Disappointingly, the hairs were so degraded that a definitive analysis was impossible.

On this expedition, the three-man team had heard the creature on several occasions. Carl Marshall heard what he described as something very akin to the sound a young gorilla makes. On another occasion, all three of them heard a high-pitched yipping/laughing sound which appeared similar to the very high-pitched laugh of the comic book character, Skeletor, of which I am vaguely aware. They also obtained a second-hand print with comparable morphology to that which was used in Kerinci National Park, on our expedition ten years ago. Moreover, they discovered a footprint which fell totally within the parameters we have established over the years for footprints of these unknown bi-ped apes.

On one occasion, they heard the high-pitched yipping from the other side of a thicket. Carl Marshall set off behind the thicket in an attempt to drive the orang-pendek forward in order for Richard Freeman to film it. However, when he reached the back of the thicket, the elusive animal ran off in the opposite direction. Carl caught a glimpse of something about a metre high and dark orange (described as being a little darker than my cat Captain Frunobulax the Magnificent - also known as Peanuts) running into the jungle. We assumed that it was probably an orang-pendek, but what happened next casts welcome doubt on that hypothesis.

The area where they were working has no tigers living there, and there were no historical reports of tiger existence. It was a long way from either of the two nearest tiger habitats. However, it was the intriguing thought of a mysterious entity said to be similar to a tiger but taller and more intelligent and somewhere between a spirit and a God that appealed. No-one in the West had heard of this entity, but found it very interesting nonetheless. Soon after their arrival in the area, they found a pug mark which Richard and Carl identified as a leopard. Several days later they found another pugmark of something which was undoubtedly a tiger. They put up trail cameras in the area hoping to get a picture of the orang-pendek. They managed to capture some photographs and a small moving film clip of a female tiger. Assuming the yet smaller pug mark was actually that of a younger tiger, then this was most certainly not the only tiger living in that supposedly 'tiger-less' area.

In doing this, they have confirmed that the animal behind the stories of the 'tiger god' in that area is most certainly a living, breathing creature. It is either a result of range expansion which our team has therefore been the first to note, or the result of a relict population of tigers in that area, something which has never been recorded. Either explanation is a fantastic result for the CFZ team. We are in the process of contacting the relevant authorities but I am sure you will understand that we will not be making the location of this sighting public knowledge. This is so we can do our best to preserve the tigers from predators and poachers.

On The Track
Our web TV show has gone from strength to strength this year. Currently, we put out two shows a week: On The Track (about half an hour in length) on Saturday afternoon's at 3.00pm and OTT Xtra (somewhere between 10 and 20 Minutes) on Wednesday evenings at 6.30pm. We have also started to put out supplemental shows, such as the one alluded to earlier, when we showed the Lake Champlain footage from September. We then interviewed Katy Elizabeth as and when it seemed appropriate for example, we did four extra shows when the Sumatra expedition returned.

I have been complaining for years that we don't get anywhere near as many viewing figures as some vacuous bint who claims to be an influencer and who has millions of people watching her ramblings about hair style tips. But I have always been aware that we are a minority interest. Furthermore, since the particular brand of Crypto espoused by the CFZ is considerably less popular than the brand of fantasy promulgated by people who should know better, and who burble on about bigfoot having the same sort of 'cloaking devices' shown on Star Trek, I suppose I cannot really complain. However, I am very pleased to be able to tell you that we have finally reached the level of viewing hours that will allow the CFZ TV YouTube channel to become monetised. Sorting all this out is a tedious exercise which Louis and I started today. I hope it will be complete by the time we reach the end of what is euphemistically known as the 'Festive Season'.

We have been investing money in new equipment and software in order to continually improve the show resulting in hopefully, ever higher viewing figures. I do assure you however, that you are never going to see me grinning vacuously at the camera and that my hair style will continue to be the train crash it always has been.

Publishing
So far this year, as well as our 12 monthly newsletters, we have published three books:

CFZ Yearbook 2022/23
ISBN 978-1-909488-65-6
The Centre for Fortean Zoology (CFZ) is a professional and scientific organisation dedicated to cryptozoology: the study of unknown animals and allied disciplines. Since 1992, we have carried out extensive research into mystery animals and animal mysteries around the globe. We produce a weekly WebTV show called On the Track (OTT), which covers cryptozoology, natural history and green issues, mixed with a little light (and often peculiar) comedy. We also operate our own publishing house, producing both magazines and books on subjects that would otherwise not see the light of day.

The Centre for Fortean Zoology Yearbook is a collection of papers and essays too long and detailed for publication in the CFZ journal, Animals & Men. With contributions from both well-known researchers and relative newcomers to the field, the yearbook provides a forum where new theories can be expounded and work on little-known cryptids discussed.

https://cfz.org.uk/2022/04/special-offer-cfz-yearbook-2022-3/

Eagle Clan Arawak Monsters - by Damen Cory
ISBN 978-1-909488-66-3
This is a remarkable book told from a unique perspective. Damon Corrie is a hereditary chief of the Eagle Clan of the Arawak Tribe based mostly in Guyana. He has made a lifelong study of

the mystery animals and animal folklore of his people, and we believe that this is the first time that these remarkable accounts have been collected together in a single volume. Moreover, the CFZ's very own Richard Freeman has added a number of appendices describing the expedition that he and Damon went on in 2007.

We heartily recommend this new volume to anybody interested in the mysteries of South America and the mythology which shapes the people who still live there.
https://cfz.org.uk/2022/05/special-offer-eagle-clan-arawak-monsters/

Invizikids - by Mike Hallowell
ISBN 978-1-909488-67-0
I remember well the day that Mike Hallowell sent me this manuscript. How impressed I was, and how disappointed I was too that he had already sold the publishing rights. 15 years on, give or take a week or two, and Mike and his original publisher have very kindly signed over the rights to us. I cannot recommend this book highly enough. It is, in my humble opinion, one of the best books that we have ever published, and, indeed, one of the best books ever written about a Fortean subject.

This is a fantastic book and centres on the imaginary friends which so many children have. I had one when I was a child, and probably so did you. Writing in 2015, eight years after this book was first published Lawrence Kutner, Ph.D. says:

"Imaginary friends are an integral part of many children's lives. They provide comfort in times of stress, companionship when they're lonely, someone to boss around when they feel powerless, and someone to blame for the broken lamp in the living room. Most important, an imaginary companion is a tool young children use to help them make sense of the adult world."

These friendships take place in a weird sort of ur-space that is neither pure imagination or actual reality (whatever that is) and are important not just because they are practically universal, but because of their implications for those of us who study the noosphere.

I have known Mike Hallowell for well over 20 years, and have always been impressed by his books and articles. He has the sort of enquiring poly mathematical mind which I admire, as he sets his talents into investigating a wide range of different arcane subjects. But this inquiry into the true nature of childhood imaginary friends may well prove to be the most important thing he has ever written.

I note with regret that this has been another year in which our journal, Animals & Men has not been published. There are simply not enough hours in the day, but I have every hope that issue 71 will be out in the next few months. However, I would be very grateful if there is anybody out there who would like to volunteer for the magazine, particularly with the news aggregation element which is a time-consuming task and one which I need someone to help with. It could probably mean that the magazine would be able to go back to something approximating its original publication schedule. I am sure with its advent to the lovely Karen, and her husband Richard - proofreader par excellence - it is going to make things in that part of the CFZ multiverse far more efficient than it is at the moment.

Animals
I look back somewhat nostalgically to the CFZ ten or twelve years ago when we had basically

fulfilled my childhood dreams of having my own private zoo. But back in those days there were a lot of people working here, and I seemed to have become a de facto uncle to half of North Devon; there were a lot of beautiful teenage girls milling about and a plethora of strapping young lads coming to see what the teenagers were doing, and I managed to put them all to work.

Now the teenagers have grown up with kids of their own, the students who used to come here for work experience are all qualified, and the only people living full time in the house are me and Graham. So, as the animals in the CFZ menagerie have passed on to pastures new (either in heaven or on earth) we have not replaced them. The aviaries have been demolished or stand empty waiting for the sort of animal rescue that the universe seems to plonk upon our doorstep more often than can be explained, but more often than pure chance. I have two functioning tanks of fish rather than the 17 I had in 2009.

Currently, the list of CFZ livestock consists of four cats, a bumptious little dog, who as you can see from this year's CFZ Christmas card, considers himself to be the most important animal or human living here. Apart from this, we still have the African leaf fish which Max donated to us 12 years ago, two axolotls, a couple of mosquito fish, both females so they won't breed! (they are practically impossible to get hold of so my breeding colony that lasted the best part of nine years will fizzle out), and Corinna's old pet hen Henny-Penny who is now about ten years old. Whether or not I end up getting more exotics depends on the way the good Lord treats my various health problems, but from where I am sitting at the moment, I think it is probably unlikely.

So, ladies & gentlemen, that is about it. I have various books which I want to publish in 2023, the first of which being a massive book on Zooform Phenomena called 'The Highest Strangeness' by Richard Freeman, which has been edited by Miss Guinevere, who has worked with him on the footnotes.

We are hoping that there will be another major expedition next year and hope to announce the details in February or March. Apart from that, we shall just keep on keeping on, as Bob Dylan once wrote.

I would like to draw your attention to this campaign, 'Kelly Kettles for Ukraine', which has been started by my brother and sister-in-law:
https://tinyurl.com/4fn5a7j9

It is a remarkable piece of practical help to the beleaguered people of that war-torn country, and I would urge you all to help in any way you can.

As always, I would like to wish you a happy Christmas and a peaceful and secure New Year.

Best wishes,
Jon Downes
Director, Centre for Fortean Zoology

STILL ON THE TRACK OF UNKNOWN ANIMALS

The Centre for Fortean Zoology, or CFZ, is a non profit-making organisation founded in 1992 with the aim of being a clearing house for information, and coordinating research into mystery animals around the world.

We also study out of place animals, rare and aberrant animal behaviour, and Zooform Phenomena; little-understood "things" that appear to be animals, but which are in fact nothing of the sort, and not even alive (at least in the way we understand the term).

Not only are we the biggest organisation of our type in the world, but - or so we like to think - we are the best. We are certainly the only truly global cryptozoological research organisation, and we carry out our investigations using a strictly scientific set of guidelines. We are expanding all the time and looking to recruit new members to help us in our research into mysterious animals and strange creatures across the globe.

Why should you join us? Because, if you are genuinely interested in trying to solve the last great mysteries of Mother Nature, there is nobody better than us with whom to do it.

We publish a journal *Animals & Men*. Each issue contains nearly 100 pages packed with news, articles, letters, research papers, field reports, and even a gossip column! The magazine is Royal Octavo in format with a full colour cover. You also have access to one of the world's largest collections of resource material dealing with cryptozoology and allied disciplines, and people from the CFZ membership regularly take part in fieldwork and expeditions around the world.

The CFZ is managed by a board of trustees, with a non-profit making trust registered with HM Government Stamp Office. The board of trustees is supported by a Permanent Directorate of full and part-time staff, and advised by a Consultancy Board of specialists - many of whom are world-renowned experts in their particular field. We have regional representatives across the UK, the USA, and many other parts of the world, and are affiliated with other organisations whose aims and protocols mirror our own.

You'll find that the people at the CFZ are friendly and approachable. We have a thriving forum on the website which is the hub of an ever-growing electronic community. You will soon find your feet. Many members of the CFZ Permanent Directorate started off as ordinary members, and now work full-time chasing monsters around the world.

Write to us, e-mail us, or telephone us. The list of future projects on the website is not exhaustive. If you have a good idea for an investigation, please tell us. We may well be able to help.

We are always looking for volunteers to join us. If you see a project that interests you, do not hesitate to get in touch with us. Under certain circumstances we can help provide funding for your trip. If you look on the future projects section of the website, you can see some of the projects that we have pencilled in for the next few years.

In 2003 and 2004 we sent three-man expeditions to Sumatra looking for Orang-Pendek - a semi-legendary bipedal ape. The same three went to Mongolia in 2005. All three members started off merely subscribers to the CFZ magazine. Next time it could be you!

We have no magic sources of income. All our funds come from donations, membership fees, and sales of our publications and merchandise. We are always looking for corporate sponsorship, and other sources of revenue. If you have any ideas for fund-raising please let us know. However, unlike other cryptozoological organisations in the past, we do not live in an intellectual ivory tower. We are not afraid to get our hands dirty, and furthermore we are not one of those organisations where the membership have to raise money so that a privileged few can go on expensive foreign trips. Our research teams, both in the UK and abroad, consist of a mixture of experienced and inexperienced personnel. We are truly a community, and work on the premise that the benefits of CFZ membership are open to all.

Reports of our investigations are published on our website as soon they are available. Preliminary reports are posted within days of the project finishing.

Each year we publish a 200 page yearbook containing research papers and expedition reports too long to be printed in the journal. We freely circulate our information to anybody who asks for it.

We have a thriving YouTube channel, CFZtv, which has well over two hundred self-made documentaries, lecture appearances, and episodes of our monthly webTV show. We have a daily online magazine, which has over a million hits each year.

From 2000—2016 we held our annual convention - the Weird Weekend. It went on hiatus because of the illness of several of the major personnel and the eventual death of one of them. But we plan to bring it back soon. It is three days of lectures, workshops, and excursions. But most importantly it is a chance for members of the CFZ to meet each other, and to talk with the members of the permanent directorate in a relaxed and informal setting and preferably with a pint of beer in one hand. Since 2006 - the Weird Weekend has been bigger and better and held in the idyllic rural location of Woolsery in North Devon.

Since relocating to North Devon in 2005 we have become ever more closely involved with other community organisations, and we hope that this trend

will continue. We have also worked closely with Police Forces across the UK as consultants for animal mutilation cases, and we intend to forge closer links with the coastguard and other community services. We want to work closely with those who regularly travel into the Bristol Channel, so that if the recent trend of exotic animal visitors to our coastal waters continues, we can be out there as soon as possible.

Apart from having been the only Fortean Zoological organisation in the world to have consistently published material on all aspects of the subject for over a decade, we have achieved the following concrete results:

• Disproved the myth relating to the headless so-called sea-serpent carcass of Durgan beach in Cornwall 1975

• Disproved the story of the 1988 puma skull of Lustleigh Cleave

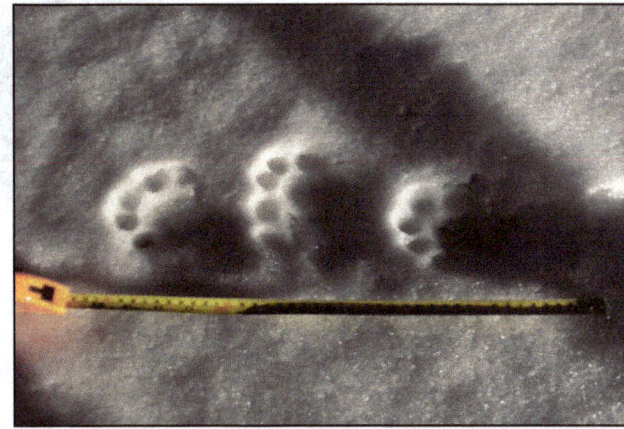

- Carried out the only in-depth research ever into the mythos of the Cornish Owlman.
- Made the first records of a tropical species of lamprey
- Made the first records of a luminous cave gnat larva in Thailand
- Discovered a possible new species of British mammal - the beech marten
- In 1994-6 carried out the first archival fortean zoological survey of Hong Kong
- In the year 2000, CFZ theories were confirmed when a new species of lizard was added to the British List
- Identified the monster of Martin Mere in Lancashire as a giant wels catfish
- Expanded the known range of Armitage's skink in the Gambia by 80%
- Obtained photographic evidence of the remains of Europe's largest known pike
- Carried out the first ever in-depth study of the ninki-nanka
- Carried out the first attempt to breed Puerto Rican cave snails in captivity
- Were the first European explorers to visit the `lost valley` in Sumatra
- Published the first ever evidence for a new tribe of pygmies in Guyana
- Published the first evidence for a new species of caiman in Guyana
- Filmed unknown creatures on a monster-haunted lake in Ireland for the first time

- Had a sighting of orang pendek in Sumatra in 2009
- Found leopard hair, subsequently identified by DNA analysis, from rural North Devon in 2010
- Brought back hairs which appear to be from an unknown primate in Sumatra
- Published some of the best evidence ever for the almasty in southern Russia

CFZ Expeditions and Investigations include:

- 1998 Puerto Rico, Florida, Mexico (Chupacabras)
- 1999 Nevada (Bigfoot)
- 2000 Thailand (Naga)
- 2002 Martin Mere (Giant catfish)
- 2002 Cleveland (Wallaby mutilation)
- 2003 Bolam Lake (BHM Reports)

- 2003 Sumatra (Orang Pendek)
- 2003 Texas (Bigfoot; giant snapping turtles)
- 2004 Sumatra (Orang Pendek; cigau, a sabre-toothed cat)
- 2004 Illinois (Black panthers; cicada swarm)
- 2004 Texas (Mystery blue dog)
- Loch Morar (Monster)
- 2004 Puerto Rico (Chupacabras; carnivorous cave snails)
- 2005 Belize (Affiliate expedition for hairy dwarfs)
- 2005 Loch Ness (Monster)
- 2005 Mongolia (Allghoi Khorkhoi aka Mongolian death worm)

- 2006 Gambia (Gambo - Gambian sea monster , Ninki Nanka and Armitage's skink
- 2006 Llangorse Lake (Giant pike, giant eels)
- 2006 Windermere (Giant eels)
- 2007 Coniston Water (Giant eels)
- 2007 Guyana (Giant anaconda, didi, water tiger)
- 2008 Russia (Almasty)
- 2009 Sumatra (Orang pendek)
- 2009 Republic of Ireland (Lake Monster)
- 2010 Texas (Blue Dogs)
- 2010 India (Mande Burung)
- 2011 Sumatra (Orang-pendek)
- 2012 Sumatra (Orang Pendek)
- 2014 Tasmania (Thylacine)
- 2015 Tasmania (Thylacine)
- 2016 Tasmania (Thylacine)
- 2017 Tasmania (Thylacine)
- 2018 Tajikistan (Gul)
- 2020 Forest of Dean (Lynx)
- 2022 Sumatra (Orang Pendek)

For details of current membership fees, current expeditions and investigations, and voluntary posts within the CFZ that need your help, please do not hesitate to contact us.

The Centre for Fortean Zoology,
Myrtle Cottage,
Woolfardisworthy,
Bideford, North Devon
EX39 5QR

Telephone 01237 431413
Fax+44 (0)7006-074-925
eMail info@cfz.org.uk

Websites:

www.cfz.org.uk
www.weirdweekend.org

THE WORLD'S WEIRDEST PUBLISHING COMPANY

HOW TO START A PUBLISHING EMPIRE

Unlike most mainstream publishers, we have a non-commercial remit, and our mission statement claims that "we publish books because they deserve to be published, not because we think that we can make money out of them". Our motto is the Latin Tag *Pro bona causa facimus* (we do it for good reason), a slogan taken from a children's book *The Case of the Silver Egg* by the late Desmond Skirrow.

WIKIPEDIA: "The first book published was in 1988. *Take this Brother may it Serve you Well* was a guide to Beatles bootlegs by Jonathan Downes. It sold quite well, but was hampered by very poor production values, being photocopied, and held together by a plastic clip binder.

In 1988 A5 clip binders were hard to get hold of, so the publishers took A4 binders and cut them in half with a hacksaw. It now reaches surprisingly high prices second hand.

The production quality improved slightly over the years, and after 1999 all the books produced were ringbound with laminated colour covers. In 2004, however, they signed an agreement with Lightning Source, and all books are now produced perfect bound, with full colour covers."

Until 2010 all our books, the majority of which are/were on the subject of mystery animals and allied disciplines, were published by `CFZ Press`, the publishing arm of the Centre for Fortean Zoology (CFZ), and we urged our readers and followers to draw a discreet veil over the books that we published that were completely off topic to the CFZ.

However, in 2010 we decided that enough was enough and launched a second imprint, `Fortean Words` which aims to cover a wide range of non animal-related esoteric subjects. Other imprints will be launched as and when we feel like it, however the basic ethos of the company remains the same: Our job is to publish books and magazines that we feel are worth publishing, whether or not they are going to sell. Money is, after all - as my dear old Mama once told me - a rather vulgar subject, and she would be rolling in her grave if she thought that her eldest son was somehow in `trade`.

Luckily, so far our tastes have turned out not to be that rarified after all, and we have sold far more books than anyone ever thought that we would, so there is a moral in there somewhere…

Jon Downes,
Woolsery, North Devon
July 2010

CFZ PRESS

CFZ Press is our flagship imprint, featuring a wide range of intelligently written and lavishly illustrated books on cryptozoology and the quirkier aspects of Natural History.

CFZ Classics is a new venture for us. There are many seminal works that are either unavailable today, or not available with the production values which we would like to see. So, following the old adage that if you want to get something done do it yourself, this is exactly what we have done.

Desiderius Erasmus Roterodamus (b. October 18th 1466, d. July 2nd 1536) said: "When I have a little money, I buy books; and if I have any left, I buy food and clothes," and we are much the same. Only, we are in the lucky position of being able to share our books with the wider world. CFZ Classics is a conduit through which we cannot just re-issue titles which we feel still have much to offer the cryptozoological and Fortean research communities of the 21st Century, but we are adding footnotes, supplementary essays, and other material where we deem it appropriate.

http://www.cfzpublishing.co.uk/

Fortean Words is a new venture for us. The F in CFZ stands for "Fortean", after the pioneering researcher into anomalous phenomena, Charles Fort. Our Fortean Words imprint covers a whole spectrum of arcane subjects from UFOs and the paranormal to folklore and urban legends. Our authors include such Fortean luminaries as Nick Redfern, Andy Roberts, and Paul Screeton. . New authors tackling new subjects will always be encouraged, and we hope that our books will continue to be as ground-breaking and popular as ever.

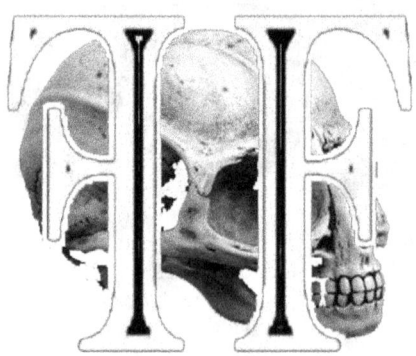

Just before Christmas 2011, we launched our third imprint, this time dedicated to - let's see if you guessed it from the title - fictional books with a Fortean or cryptozoological theme. We have published a few fictional books in the past, but now think that because of our rising reputation as publishers of quality Forteana, that a dedicated fiction imprint was the order of the day.

http://www.cfzpublishing.co.uk/

www.ingramcontent.com/pod-product-compliance
Lightning Source LLC
Chambersburg PA
CBHW072134270326
41931CB00010B/1756